TOTAL
MANUFACTURING
ASSURANCE

QUALITY AND RELIABILITY

A Series Edited by

Edward G. Schilling

Center for Quality and Applied Statistics
Rochester Institute of Technology
Rochester, New York

TOTAL MANUFACTURING ASSURANCE

CONTROLLING PRODUCT RELIABILITY, SAFETY, AND QUALITY

DOUGLAS C. BRAUER
Panduit Corp.
Tinley Park, Illinois

JOHN CESARONE
University of Illinois
Chicago, Illinois

Marcel Dekker, Inc.   New York • Basel • Hong Kong
ASQC Quality Press   Milwaukee

ISBN 0-8247-8441-3

This book is printed on acid-free paper.

ASQC Quality Press
310 West Wisconsin Avenue, Milwaukee, Wisconsin 53203
Marcel Dekker, Inc.
270 Madison Avenue, New York, New York 10016
Current printing (last digit):
10 9 8 7 6 5 4 3 2 1

PRINTED IN THE UNITED STATES OF AMERICA

About the Series

The genesis of modern methods of quality and reliability will be found in a simple memo dated May 16, 1924, in which Walter A. Shewhart proposed the control chart for the analysis of inspection data. This led to a broadening of the concept of inspection from emphasis on detection and correction of defective material to control of quality through analysis and prevention of quality problems. Subsequent concern for product performance in the hands of the user stimulated development of the systems and techniques of reliability. Emphasis on the consumer as the ultimate judge of quality serves as the catalyst to bring about the integration of the methodology of quality with that of reliability. Thus, the innovations that came out of the control chart spawned a philosophy of control of quality and reliability that has come to include not only the methodology of the statistical sciences and engineering, but also the use of appropriate management methods together with various motivational procedures in a concerted effort dedicated to quality improvement.

This series is intended to provide a vehicle to foster interaction of the elements of the modern approach to quality, including statistical applications, quality and reliability engineering, management, and motivational aspects. It is a forum in which the subject matter of these various areas can be brought together to allow for effective integration of appropriate techniques. This will promote the true benefit of each, which can be achieved only through their interaction. In this sense, the whole of quality and reliability is greater than the sum of its parts, as each elements augments the others.

The contributors to this series have been encouraged to discuss fundamental concepts as well as methodology, technology, and procedures at the leading edge of the discipline. Thus, new concepts are placed in proper perspective in these evolving disciplines. The series is intended for those in manufacturing, engineering, and marketing and management, as well as the consuming public, all of whom have an interest and stake in the improvement and maintenance of quality and reliability in the products and services that are the lifeblood of the economic system.

The modern approach to quality and reliability concerns excellence: excellence when the product is designed, excellence when the product is made, excellence as the product is used, and excellence throughout its lifetime. But excellence does not result without effort, and products and services of superior quality and reliability require an appropriate combination of statistical, engineering, management, and motivational effort. This effort can be directed for maximum benefit only in light of timely knowledge of approaches and methods that have been developed and are available in these areas of expertise. Within the volumes of this series, the reader will find the means to create, control, correct, and improve quality and reliability in ways that are cost effective, that enhance productivity, and that create a motivational atmosphere that is harmonious and constructive. It is dedicated to that end and to the readers whose study of quality and reliability will lead to greater understanding of their products, their processes, their workplaces, and themselves.

Edward G. Schilling

Preface

Total Manufacturing Assurance (TMA) is the attainment of the ability to ensure products are manufactured in a manner that enhances reliability, safety, and quality. These performance attributes lead to successful product commercialization and profitability. Thus, TMA is a necessary and fundamental objective of the corporate management mindset, as well as of the overall corporate culture.

TMA, as a corporate initiative, is embraceable by every person involved in the manufacturing environment. This includes (but is not limited to) corporate management, engineering management and engineers, and production management and workers. However, it takes a well-planned and coordinated effort to achieve TMA.

The purpose of this book is to provide insight into the key concepts, tools, and techniques that are integral aspects of TMA. There is much emotion in addressing TMA because it is real and tangible and, as a corporate issue, cannot be avoided. If no, or incomplete, attention is paid to manufacturing assurance, numerous undesirable situations will arise. Manufacturing processes will be faulty (because of poor or incomplete task operation), production lines will be faulty (because of excessive downtime and rework), corporate morale will be low (because of management blaming the work force and the work force blaming management), and, consequently, the end product will be faulty. The result is obvious: either the product or the company fails. Neither of these parallels corporate goals.

It is important to emphasize that the purview of TMA's contributing elements address management, engineering, marketing, and the work force. Each of these areas depends on the others to be successful. Therefore, it is impossible to achieve TMA without cooperation and communication between these groups.

A goal of this book is to illustrate the bond that must be established between these and all corporate groups, and the input and leadership that each group must provide. This relationship is supported by the various qualitative and quantitative tools and techniques presented. Open communication and appropriate technical aids effect a straightforward path to expedite attainment of TMA.

In simple terms, the straightforward path must balance four perspectives. TMA does this. From a management perspective, TMA enhances cost-effectiveness. From an engineering perspective, TMA optimizes the overall manufacturing system and output products as designed. From a marketing perspective, TMA provides a product that is desirable to the consumer. Finally, from a work force perspective, TMA promotes practicality.

The book is divided into four parts: (1) Introduction, (2) Strategic Manufacturing Management, (3) Manufacturing System Control, and (4) System Improvement Monitoring. This enables the reader to focus on a particular aspect of TMA.

The Introduction section consists of two chapters. Chapter 1 provides a brief overview of manufacturing as it existed yesterday, exists today, and is envisioned tomorrow. Discussion also focuses on some philosophical remarks regarding manufacturing management. Chapter 2 introduces and defines the three major elements of TMA; that is, strategic manufacturing management, manufacturing system control, and system improvement monitoring.

The Strategic Manufacturing Management section consists of two chapters. Chapter 3 addresses manufacturing planning from both a strategic and a tactical standpoint. Chapter 4 addresses management control. It discusses the necessity for an integrated approach involving all participating corporate activities to attain TMA. Also discussed are motivation, qualitative and quantitative tools, and expert systems. The latter is enthusiastically included in this book to illustrate its power and usefulness.

The Manufacturing System Control section consists of three chapters. Chapter 5 discusses several key issues impacting the definition of a manufacturing system, including layout/flow, materials handling, automation, and process simulation. Chapter 6 provides insight into controlling product degradation with regard to reliability, safety, and quality as a function of the overall manufacturing process. Chapter 7 addresses system maintenance from the standpoint of optimizing manufacturing process availability. The reliability-centered maintenance approach is presented as a means to achieving this optimization.

The System Improvement Monitoring section consists of two chapters. Chapter 8 addresses data system planning and Chapter 9 addresses data recording and feedback. Both these chapters highlight issues increasingly important to corporate management.

The book's intended audience is management, engineering, and others intimately involved in the manufacturing process. It is also intended to serve as an instructional resource for engineering management, mechanical and industrial engineering, and business students (both university advanced undergraduate and graduate levels). The material presented addresses management and technical topics in the level of detail necessary for their practical understanding and implementation to attain TMA.

The authors thank several persons for their direct and indirect significant contributions to the writing of this book. These include Ms. Jeanine Lau of ASQC; Professor Floyd G. Miller of the University of Illinois at Chicago; Mr. Sidney Bass of Sid Bass and Associates; and Mr. Ronald T. Anderson and Dr. Charles D. Henry of Reliability Technology Associates. Also, Mr. Greg D. Brauer of Fireman's Fund Insurance Company provided input regarding safety engineering management.

A special acknowledgment goes to Ms. Laurie Brauer for her work in preparing all graphic artwork. Thanks for a job well done.

Douglas C. Brauer
John Cesarone

Contents

Part One

INTRODUCTION

Never shrink from doing anything which your business calls you to do. The man who is above his business, may one day find his business above him.

—Drew

1

The World of Manufacturing

Many of the innovative manufacturing concepts in practice throughout the world today are directly attributable to American ingenuity. Such innovative concepts find their roots in the manufacturing evolution that has occurred as an integral part of American history. This chapter highlights some of the historical events that preceded our current manufacturing state of the art and will no doubt inspire future achievements.

This chapter presents a brief overview of some key historical events and persons. We then shift our attention from history to current trends and state-of-the-art technology. Finally, we take up manufacturing enhancements that will be fully realized in the near future. This sequential presentation is intended to create an awareness of the impact that manufacturing has on overall corporate direction. The chapter also includes a section presenting some philosophical ground rules for all levels of corporate management.

I. INTRODUCTION

Manufacturing can be defined as the process of making a product from raw material in accordance with an organized plan. A piece of cake. Anybody can do it. I can do manufacturing blindfolded and with one hand tied behind my back. Just give me a little raw material and I'll show you.

All too often this seems to be the carefree attitude taken. If we have some raw material, a couple of machines, and a vague manufacturing plan, then it is only a matter of time before the products start rolling out and the money starts rolling in. Obviously, there is much more involved in manufacturing than merely shotgunning hardware out the plant door.

Three fundamental concerns of a manufacturing organization must be integrated to ensure success: (1) corporate management (which includes marketing, sales, finance, etc.), (2) engineering, and (3) production. Each of these concerns centers on people and each provides key contributions in the overall manufacturing effort. Yet without the other two, each is useless. A team effort is necessary for corporate success.

The corporate structure and its ability to be successful is analogous to a needle and thread. If there is a common thread binding each member of the corporation together, the probability for success is enhanced. The thread itself is the organizational management approach and the corporate culture, or attitude, instilled in the team members.

A sharp needle must lead the thread. The needle is the fundamental corporate direction pointed by the chief executive officer or president based on the corporate strategic business plan. The size of the needle is important; it must fit the company. If the hole at the end of the needle is too small, it is difficult to get a thread through, and it is also tough for the needle to lead the thread if the two are separated. Likewise, if the hole is too large, there is a lot of space in which the thread can move around, and it becomes hard to lead and control the position of the thread.

Assuming that the needle gets threaded, the immediately following part of the thread is the rest of the corporate management staff. In this area the thread is doubled; the length of this doubled area depends on how top-heavy the corporation's management is. A smaller doubled area indicates greater management efficiency. Completing the thread is engineering management and engineering, production management, the production personnel, and others.

To ensure corporate success, there must be a knot at the end of the thread. The size of the knot is a function of the corporate strategic business plan's robustness. If the knot is too small, or does not exist, the thread slips when pulled to manufacture products. If the thread is pulled too hard, it breaks or the knot slips through the hole made in the marketplace. The result is a severe discontinuity in management-team member relations.

Does all this mean that management must learn to sew? Certainly not. It means that the corporation is a team. The manufacture of products is a result of many people working together to achieve a common goal.

Obviously, there must be a hierarchy of management to ensure that products do get manufactured correctly, efficiently, and on time. However, the full potential of any corporation cannot be realized if there is a lack of mutual respect and caring between persons positioned on different levels of the corporate ladder. A strong binding thread must be in place.

II. YESTERDAY

The current state of manufacturing is a result of much patient learning and heuristic applications of innovative concepts. This learning has by no means stopped and it must not. Global and national markets are becoming increasingly competitive. Aside from the issue of cost, there is an ever-increasing consumer awareness of the right to high-quality products and services.

By taking a quick look at history, it becomes clear that much progress has been made. For the most part corporations have evolved from a neanderthal management mindset to a more progressive, understanding, and compassionate style. (Unfortunately, this not universally true.)

A good place to pick up the history of manufacturing is the industrial revolution. Industrialization resulted in an accelerated use of, and need for, mechanized manufacturing systems. This era may very well have witnessed the birth of the manufacturing engineer.

The revolution itself began in England in the early eighteenth century. It rapidly spread to other parts of the world during the remainder of the eighteenth century and into the nineteenth century. Although the industrial revolution gained much worldwide momentum, it did not affect all countries. Even today, there are many areas in the world that have remained essentially untouched by mechanized manufacturing system technology.

During the latter part of the 1800s, several countries emerged as major industrial leaders. England, the United States, Germany, Italy, and Japan became the most important industrial nations.

One key beneficiary of the great strides being made in mechanization was agriculture. Advances in engineering also made it possible to improve and employ innovation in manufacturing processes. Machines and equipment began to replace inefficient systems driven by manually created motive power.

A classic example is the steam engine. Steam power characterized the nineteenth century at its apex. Other sources of power—including electricity, petroleum, and nuclear power—characterized the continuing industrial revolution of the twentieth century.

The industrial revolution had far-reaching socioeconomic consequences. These varied in detail from country to country and are still present in countries undergoing technological development. The most obvious and direct consequence was a tremendous increase in the production of material goods. For example, mechanization enabled textile plants in England to produce billions of yards of cotton cloth a year. This increase in production was many times more than people could produce working with old-fashioned spinning wheels and hand looms.

By 1900 the United States had become the greatest manufacturing nation in the world. Technological skill, government policies, business leadership, natural resources, labor supply, and large markets all contributed to the rapid progress made. The combination of these advantages enabled the United States to become the world leader in industrial productivity.

While the total output of industry in the United States was increasing, the number of individual plants was decreasing. As plants grew fewer and (on the average) larger, so did the business firms that operated them. Before the Civil War, the typical company was owned and run by an individual or partners, though a number of corporations existed. After the war, corporations became larger and more numerous.

For the owners of a company, large size provided certain advantages. It enabled the company to lower its costs through mass purchasing and mass production. Often, it also served as a tool to weaken or eliminate competitors, thus enabling the company to maintain or even raise prices. For consumers, large businesses were not necessarily advantageous. To the extent that low costs led to low prices, the consumer benefited. However, in many instances the benefits never came to fruition.

Significant events in the United States during this period created fundamental changes in demographics. People moved to large cities, medical advances caused people to live longer, and new means of transportation allowed people to disperse around the country. Also at this time a great insurgence of immigrants occurred.

A key side effect of these technological and economic changes was the formation of unions, or organized labor groups. This, in turn, led to the formation of the classic industrial conflict between corporations and unions.

As corporations become the owners of industries, organized labor became necessary to offset corporate power and consequent employee abuses. The unions implemented strikes and boycotts. Basic demands included improved working conditions, recognition of unions as bargaining agents, and the passing of federal laws to regulate various industries.

These early years were shaped by many well-known business persons. People such as Cyrus McCormick (reaper works), Charles Goodyear (vulcanized rubber), Isaac Singer (sewing machine), Samuel Colt (gun mass production), John Rockefeller (oil), Andrew Carnegie (steel), Thomas Edison (electricity/inventions), George Westinghouse (electricity), J. Pierpont Morgan (finance/banking), Ransom Olds (automobile), and Henry Ford (automobile) provided key contributions to our manufacturing management and engineering knowledge and growth. However, as manufacturing capability, efficiency, and productivity increased, so also did the indifference of corporate management to the fundamental needs of the work force; that is, respect, honesty, and caring.

III. TODAY

Without a doubt the list of contributors to manufacturing management, organization, and engineering innovation fills several pages. Their contributions enabled major strides to be made in technological and management advancements. Corporations such as Ford, Motorola, General Motors, International Business Machines, Xerox, Kodak, and many others continue to push the state of the art in both automated manufacturing and management concepts.

The automated manufacturing capabilities of today (e.g., computer-integrated manufacturing, flexible manufacturing systems) are far superior to the capabilities available to Henry Ford. But many of the manufacturing philosophies in use around the world today were seeded by people of that era. For example, in the early 1900s, the just-in-time approach to manufacturing was used in a crude form by Ford; Statistical Process Control was defined by Shewart; and Design of Experiments was developed by Fisher. These manufacturing management/control tools were the subject of much discussion and use during the 1980s and will be well into the 1990s.

With the rediscovery annd use of many innovative manufacturing ideas that were ahead of their time, to increase efficiency, productivity, and profitability, corporations are taking to heart the right of the consumer to get reliable, safe, and quality products. This perhaps was a hard rediscovery and admission on the part of corporations in the United States. It took a slap in the face from Japanese corporations to wake up the quality-unconscious corporations around the world, particularly those in the United States. Names such as Demming, Juran, Crosby, Feigenbaum, Taguchi, and Shainin identify some of the more well-known quality gurus. Their services are in high demand and their ideas are commonplacc in quality discussions and activities around the world.

IV. TOMORROW

Where do we go from here? There is little chance to escape the continuing advancements in electronics that make automated manufacturing a reality. Computers play a major role in manufacturing today, and their role will broaden with advances in the development of artificial intelligence. Artificial intelligence (AI) is a field of computer science that focuses on using computer technology to emulate the behavior of humans in solving problems. The use of AI magnifies the need for both a comprehensive data system and a supporting data recording and feedback system.

There are currently five primary applications of AI: (1) pattern-matching, (2) machine learning, (3) robotics, (4) natural language processing, and (5) expert systems. Of special interest is the use of expert systems in all organizational levels of a corporation. Expert systems imitate human experts by giving advice. These systems draw upon their own store of knowledge and also request information specific to the problem at hand. The knowledge is commonly maintained as a set of rules, and the specific information is provided by the person seeking advice. The mechanism of the expert system that combines this knowledge and information to make decisions is called the *inference engine.* Many expert system shells are commercially available. These shells provide the inference engine to drive the programs' ability to make decisions from its knowledge base.

Expert system technology will increasingly become an integral part of manufacturing system design and implementation. Computer-integrated programmable logic controllers will more effectively monitor and gather process information directly. Upon the identification of a problem, the expert system will define the corrective action necessary to maintain process control. Ultimately, the expert system will make intelligent corrective action

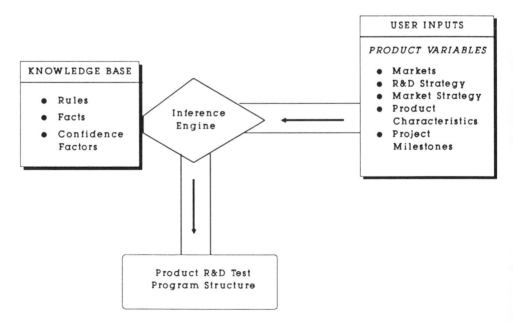

Figure 1   Conceptual Expert System

decisions on its own using the information it receives from the programmable logic controller. The expert decisions then will be relayed back to the controller, which, in turn, will implement the decisions made.

The use of this technology will not be limited to manufacturing, but also will be applied to aid corporate management in making decisions. Its use will enhance the soundness and correctness of management decisions by looking at the same rules to make a decision as would a corporation's chief executive officer or president. But in applying these same rules, or knowledge, it will give a faulty answer if its rules are faulty.

Expert systems are already being used; for example, to assist in making maintenance decisions. Other applications include product development project planning, such as that conceptualized in Figure 1.

V. POTPOURRI

Before diving into the main body of this text, here is some management food for thought.

1. Don't tolerate garbage. There is absolutely no reason to put up with mediocre or poor work and manufactured products.
2. Preact, don't react. Do not wait until your back is against the wall before taking action.
3. Enjoy what you do. Most of one's life is spent on the job. So have fun. If it is impossible to have fun, it may be time to look for some new challenges. Don't make other people miserable just because you are.
4. Maintain a sense of professionalism. Do the best job possible and continue to look for ambitious new challenges and goals. Above all, be ethical, honest, respectful, and caring toward your fellow corporate team members.
5. Don't be afraid of change. By dynamic, open-minded, and objective. Avoid falling prey to management and manufacturing fads. Never allow yourself to fall into a mode of complacency.
6. Keep your eyes on the whole picture. Work to integrate your corporate efforts with others: don't isolate your vision to one element of the corporation. Communicate effectively with people and eliminate interdepartmental walls.

2

Total Manufacturing Assurance

Total manufacturing assurance (TMA) is the focus of this book. Accordingly, this chapter defines TMA. It lets us catch our breath before jumping into the collection of tools, techniques, and philosophies that support the achievement of TMA.

This chapter provides an overview of TMA's three integral elements and the major activities of each. The first element is strategic manufacturing management. The second element is manufacturing system control. The third element is system improvement monitoring.

I. INTRODUCTION

Total manufacturing assurance (TMA) is the ability to ensure that products are manufactured in a manner that enhances reliability, safety, and quality. TMA must be the objective of corporations seeking the benefits of market share leadership and high, long-term profitability. Attaining TMA is a function of three integral elements: (1) strategically planning manufacturing management activities, (2) controlling the manufacturing system, and (3) monitoring the manufacturing system for continual improvement.

These three elements, in themselves, sound pretty simple. However, TMA requires knowledgeable foresight on the part of corporate management if it is to make persistent and comprehensive efforts in planning, implementing, and monitoring manufacturing directives.

There are many well-founded management concepts documented. Total quality control, total quality management, total preventive maintenance, Juran's quality trilogy, Deming's 14 points for management to improve productivity, Crosby's 14-step quality improvement program, and thriving on chaos are but a few. Each has its merits and each undoubtedly has numerous application success stories. It is terrific to see specialized, innovative ideas created and applied.

TMA requires looking at the whole corporate picture, not just one segment of activity such as quality control. All corporate functions and activities support the ultimate objective of profitable manufacturing and product commercialization.

Figure 1 illustrates the elements, and major activities therein, of TMA. Notice that these elements are sequential and progressive. The first element

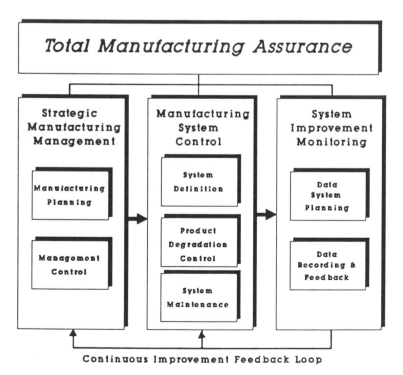

Figure 1   TMA Network

(strategic manufacturing management) feeds into the second element (manufacturing system control), which feeds into the third element (system improvement monitoring). For TMA, these activities must be close-looped. That is, the third element feeds back into elements one and two. This allows continual improvement in the TMA process.

Together, these major elements form a network to fuel the TMA process. Let's think about these two words for a moment. TMA is a process; not a program. Programs are generally activities that are planned to end upon completing a defined objective. Processes are dynamic entities that go on forever and are updated to reflect current innovative methods and concepts.

Again, we emphasize that TMA is a process. If the corporate goal is to attain TMA, and the effort that made it possible is then reduced or abandoned, you can bet your last dollar that TMA is not going to be maintained for very long. Every gain is lost and every market competitor is smiling from ear to ear.

Obviously, it is not good news if market competitors are smiling. The way to make them frown is to define and implement, and then revise as necessary, the corporate management and manufacturing objectives for launching a product into successful commercialization. This requires integrating and coordinating all of the groups (and their functions) in the company. This includes all levels of management, engineering, marketing, and production personnel. Each group works with the others and builds on the incremental corporate strides toward TMA.

The bottom line is that the TMA process leads to a shifting of many management and manufacturing assurance issues into lower cost, lower risk directives. This sets the stage for more effective and well-focused product manufacturing, which leads to a reduction in corporate waste and productivity losses.

This enables us to define the three Cs of TMA:

1. *Create* vision. Only corporations that have the ability to identify future opportunities are likely to experience long-term success and survival. Know where you will be in the future.
2. *Control* the manufacturing system. Detour any possibility of being a slave to what a manufacturing system provides. Always make the system provide high reliability, safety, and quality.
3. *Critique* performance. Know what you did yesterday and compare it to what is happening today and what is required tomorrow.

The remainder of this chapter broadly defines each element of the TMA process and the specific activities required for each. Keep in mind that the overall management of these elements must emphasize the consolidation of the TMA activities. Consolidation meaning establishing a team-oriented relationship among all corporate groups. Isolation of activities and groups is not a part of the TMA process. Remember: consolidation, not isolation.

II. STRATEGIC MANUFACTURING MANAGEMENT

Strategic manufacturing management consists of two key activities: manufacturing planning and management control.

Manufacturing planning involves establishing the direction of the corporation in the near term and the long term. By examining internal and external environmental conditions, current and projected, a corporate-wide strategy can be defined and implemented. Obviously, attaining and maintaining TMA are integral considerations in the manufacturing, as well as the business, strategic segments of the overall corporate strategy. However, in addition to defining the manufacturing strategy, the tactical implementation of this strategy must be given careful consideration, both in terms of its impact on capital investment and from the corporate culture standpoint.

Management control addresses the means by which the corporate strides toward TMA are achieved. Numerous management techniques, tools, and concepts are available as aids and should be used open-mindedly. If, for some reason, a management control technique fails to work, it should be replaced or revised. The hardest task is to change bad management habits and make an honest effort to improve worker productivity and efficiency.

III. MANUFACTURING SYSTEM CONTROL

Manufacturing system control consists of three key activities. These are system definition, product degradation control, and system maintenance.

System definition addresses the implementation of a rational method for operating the manufacturing system(s). Efficient and effective manufacturing systems are not an accident. Through proper planning and evaluation of alternatives, it is possible to put the optimal system in place. The optimal system is the design that maximizes manufacturing assurance.

Product degradation control addresses the product as it is manufactured. A product is designed to provide various performance capabilities. However, the manufactured product often exhibits performance levels lower than that designed in, particularly with regard to reliability, safety, and quality. Care must be taken to ensure that the manufacturing system is designed to minimize the degradation that will occur.

System maintenance is a critical part of the manufacturing system. Not only must system availability be maximized, but if the system is not properly maintained, product degradation is accelerated. The maintenance program defined for a manufacturing system must be considered as important as the product and system design themselves.

IV. SYSTEM IMPROVEMENT MONITORING

System improvement monitoring consists of two key activities: (1) data system planning, and (2) data recording and feedback.

Data system planning addresses the breadth of data and information required to assess the current state of TMA. Common issues that arise include what data should be collected and how to collect it. Ideally, all information sources automatically feed into a central data system. Numerous technologies exist to make this happen. An important ground rule to keep in mind is that one should collect only data and information that is practically useful, is evaluated, and is reacted to.

Data recording and feedback provide the means for collecting and evaluating pertinent data and information. This is achieved by having a closed-loop system in place that records data, looks for trends, and follows through with a proper response to highlighted issues. This is particularly important in regard to internally and externally identified problems that endanger market performance. Such problems require timely evaluation and corrective action, as appropriate.

V. SUMMARY

Now we have an understanding of what TMA is all about. It focuses on the product and on the corresponding manufacturing system(s). This focus begins with strategic manufacturing management, manufacturing system control, and system improvement monitoring.

All organizations in the corporation must participate in this focus and be committed to achieving TMA. TMA cannot be attained by going through the motions and pretending things are better. We end up fooling nobody, not even ourselves.

Remember the three C's of the TMA process. *Create* a vision of where you need to be in the future. *Control* your products and manufacturing system(s) to give you what is needed. Finally, *critique* your performance in meeting and exceeding your needs.

Part Two

STRATEGIC MANUFACTURING MANAGEMENT

There are in business three things necessary: knowledge, temper, and time.

—Feltham

3

Manufacturing Planning

An essential part of any activity is timely and comprehensive planning. Accordingly, this chapter addresses the two primary elements of manufacturing planning.

The first planning element is strategy. Strategy focuses on the business management planning issues. The second planning element is tactics. Tactics focuses on the engineering management and technology planning issues.

I. INTRODUCTION

A fundamental task in any business is planning. Without a disciplined planning effort, it is impossible for a corporation to respond to business opportunities and obstacles effectively and efficiently. The result is knee-jerk reactions to the day-to-day dynamics of our global economies. Clearly, what is necessary is a business strategic plan that enables a corporation to direct itself toward maximizing profitability.

A business strategic plan is a compilation of several strategic components. These components generally include manufacturing, finance, management, marketing, and research and development. Each requires its own strategic plan that is compatible with the strategic objectives of all other components and of the corporation as a whole.

This chapter addresses the manufacturing component of the overall corporate strategy. Accordingly, the focus is on defining and implementing a TMA-based strategy to control product reliability, safety, and quality. The goals of a manufacturing strategy are to eliminate product performance losses during manufacture, gain customer confidence and preference, capture market share, and enhance corporate profitability over the long term.

II. STRATEGIC PLANNING

Strategic planning involves several fundamental tasks: strategy formulation, strategy implementation, and evaluation and control [1]. Strategy formulation is the development of long-range plans to deal effectively with environmental opportunities and threats in light of corporate strengths and weaknesses. This consists of defining the corporate mission, specifying achievable objectives, developing strategies, and setting policy guidelines.

Strategy implementation is the process of putting strategies and policies into action through the development of various programs, budgets, and procedures. This is often referred to as operational planning, since it is concerned with day-to-day resource allocation problems. Finally, evaluation and control include the process of monitoring corporate strategic activities and results. From this evaluation, an assessment is made of actual performance relative to desired performance.

Strategic manufacturing planning is a necessary and complex activity. We have to formulate our strategy, implement it, and then monitor it to see if it works to a satisfactory level. But before a company can perform the latter two activities, it is necessary to have in place a comprehensive and complete strategic manufacturing plan.

A *strategic manufacturing plan* implies several key ideas. First, a corporate-wide strategy is defined based on input from some of the most innovative, visionary, and pragmatic minds in the company. Second, the focus is on manufacturing, and very likely some specific product line(s). Third, a detailed plan is developed reflecting realistic goals, objectives, timing milestones, and cost estimates.

The functional elements of a strategic manufacturing plan are depicted in Figure 1. Two major inputs that impact the planning process are illustrated: the business and the tactical strategic considerations. (Tactical strategic considerations are addressed later in this chapter.)

A third input is also depicted. This input, often overlooked, is the product itself. In addition to traditional performance parameters addressed in design, reliability, safety, and quality, there are critical attributes that can make or break the ability to gain customer preference and market share.

Rational manufacturing decisions are based on the defined strategic issues surrounding the corporation as a whole and, particularly, the product(s) to

Figure 1   Inputs to Manufacturing Strategy

be manufactured. Several key issues include market share potential, probability of success, and return on investment (ROI).

The market share potential for a given product varies with the growth characteristics of the specific market to be entered (assuming that successful product technology transfer occurs). Market share potential directly affects the probability of commercial success and the ROI that may be seen. Figure 2 depicts the relationship between ROI and probability of commercial success as a function of market share potential.

These strategic issues are a direct reflection of the consumers' impression of product reliability, safety, and quality. Therefore, it makes sense to develop and implement a strategic manufacturing plan that enhances product commercialization success potential and ROI.

Baseline product reliability, safety, and quality requirements are determined from identification of the strategic issues. The significance of these product attributes is evidenced in the extensive profit impact of market strategy (PIMS) data base [2]. The PIMS data implies that the definition of product reliability, safety, and quality requirements is strongly dictated by market issues. This is extended to emphasize their driving position influence in achieving high market share, ROI, and successful commercialization.

A key influence in structuring the manufacturing plan is the nature of the market to be entered. As shown in Figure 2, whether market pull or market push exists strongly affects the required product reliability, safety, and quality. Market pull indicates the situation where the product is highly desired by the consumer and will be readily received in filling a market void. Market push indicates the situation where the product is not highly desired by the consumer and, consequently, must be made desirable to the consumer.

We stated earlier that for a product to be successful it must be designed with the proper levels of reliability, safety, and quality. Accordingly, the manufacturing strategy must reflect, as one of its primary objectives, the

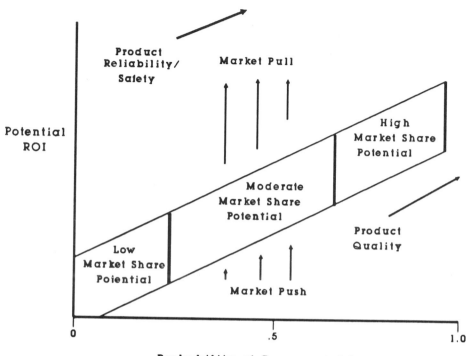

Figure 2   Product ROI/Success Relationship

need to control and maintain these inherent product performance attributes. The customer should get the product as designed, and not some degraded version of the original design.

A sound framework for adequately addressing manufacturing strategy issues is the *strategic market management* approach [3]. Strategic market management is designed to help precipitate and make strategic decisions. A strategic decision involves the creation, change, or retention of a strategy. An important role is to precipitate, as well as make, strategic decisions. Figure 3 shows a pyramid of the analyses that provide the input to strategy development and the strategic decisions that are the ultimate output. As illustrated in the figure, two fundamental analysis areas aid strategic formulation: external analysis and internal analysis.

External analysis consists of examining variables that exist outside the corporation, which typically are not within the short-term control of management. Specific analyses performed include (1) customer analysis to identify

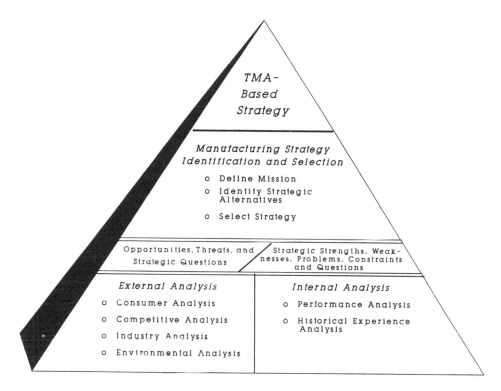

Figure 3 Strategic Planning Pyramid

customer segments, motivations, and needs; (2) competitive analysis to identify other companies' strategies, objectives, cultures, costs, strengths, and
weaknesses; (3) industry analysis to identify current size, potential growth,
available distribution systems, entry barriers, and success factors; and (4) environmental analysis to identify factors of existing technologies, government
controls, cultures, demographics, and economics.

Internal analysis consists of examining variables within the corporation.
These variables also are not typically within the short-term control of management. Specific analyses include (1) performance analysis to identify current and projected return on assets, market share, product value and performance, cost, new product activity, corporate culture, productivity, and
product portfolios; and (2) historical experience analysis to identify past
and current strategies, problem areas, organizational capabilities and constraints, financial resources available, and corporate flexibility.

Once a strategic manufacturing plan is defined, it should be evaluated
relative to several issues. These include (1) internal consistency, (2) consistency

with the environment, (3) capabilities of available resources, (4) acceptable level of risk, (5) appropriate time horizon, and (6) overall feasibility.

Internal consistency refers to the cumulative impact of a strategy on organizational objectives. Consistency with the environment refers to strategy compatibility with what is going on outside the organization. Capabilities of available resources refers to what an organization has to help it achieve its objectives. Balance must be achieved between strategies and resources. The acceptable level of risk refers to the uncertainty of strategies. Different strategies involve different risks (both internal and external), and management must establish its risk preferences. As time horizons increase, risk increases, particularly in unstable markets. The appropriate time horizon refers to the time period over which the strategies are be followed. The overall feasibility refers to the workability of strategies on a practical level.

For a manufacturing strategic plan to be successful, it must be compatible with all the strategic components involved in the corporation and with the particular product strategy. Typically, there are four product strategies: (1) market leader, (2) follow-the-leader, (3) applications-oriented, and (4) production efficient [4].

1. *Market leader strategy.* An intense product research and development (R/D) and engineering effort is implemented. Technical leadership is exerted through large financial investment(s). This strategy involves a high degree of risk, and usually only large organizations with substantial resources are able to adopt this posture, because the corporation must be able to absorb financial mistakes.

2. *Follow-the-leader strategy.* This strategy involves minor product research but a strong development and engineering effort. It requires a rapid technical response to new products developed by market leaders. Manufacturing and marketing expertise are also vital to success. The market is generally entered with a competing product during the growth stage of the product life cycle.

3. *Applications-oriented strategy.* This strategy involves minor product R/D but a strong engineering effort. The focus is on developing product modifications to serve a particular, specialized, or limited market segment. The market is generally entered during the maturity stage of the product life cycle.

4. *Production-efficient strategy.* This strategy involves little product R/D and engineering effort but a strong manufacturing effort. It is based on achieving superior manufacturing efficiency and cost control. Competition in price and delivery are paramount. The market is generally entered during the maturity stage of the product life cycle.

A preference for manufacturing strategies with greater flexibility is desirable. Strategies that permit corrective action and adjustment are better than those that cannot be changed once they are implemented.

A. Product Assurance Strategy

1. Reliability, Safety, and Quality Issues

The product reliability, safety, and quality requirements are a key part of the overall manufacturing plan. As stated earlier, this is supported by the PIMS data base. These performance parameters provide a basis for defining the product assurance strategy necessary as part of the overall manufacturing strategy.

Developing a product assurance strategic plan involves balancing many interrelated variables and factors. These include

1. Establishing quantitative product performance goals (mean time between failure, mean time to repair, service life, defectivity level, and so forth) based on customer requirements
2. Defining and implementing an effective management control program to guide the product development effort and provide timely outputs consistent with major design and program decision points
3. Performing and implementing analyses and quality checks, audits, and controls to ensure that all test, inspection, and screen data are complete and acceptable
4. Performing a fully coordinated product development program that emphasizes failure analysis and corrective action and provides growth and verification of specified design requirements
5. Establishing a set of documentation through which conformance to the product assurance strategy is assessed and tracked
6. Exposing design deficiencies and initiating corrective action in a timely manner
7. Improving product operational availability to enhance the probability of commercial success
8. Reducing the need for maintenance and logistic support (reduce life-cycle costs)
9. Identifying, evaluating, and eliminating design hazards and risks
10. Applying historical data, including lessons learned from other development projects, to the design process
11. Minimizing unforeseen impacts on overall project cost and schedule

The product assurance strategy provides an organized method of ensuring that the above factors (and others) are adequately considered during a product development program. With a strategy developed and documented, it is possible to initiate a properly planned program. Such a program ensures that critical project decisions involving significant investment of resources are keyed to the achievement of specific reliability, safety, and quality requirements, as well as other performance requirements.

The specific program objectives depend on the uniqueness of the technology under consideration, the nature of the product, the customer, and other factors as appropriate. Merely defining a product assurance strategy orients product development and, ultimately, manufacture toward a practical, serviceable, and affordable commercial product.

A product assurance strategy promotes and enables trade-off analyses between design and test engineering fucntions. Once implemented, it avoids duplication of effort and minimizes the probability of omitting essential elements. Always remember that there is an intimate relationship between the total product assurance effort and the TMA-based strategic plan.

III. TACTICAL PLANNING

Tactical manufacturing planning encompasses those planning activities that follow the strategic issues discussed in the previous section. This does not imply that strategic and tactical issues are completely separate aspects of the manufacturing enterprise. Rather, in the true spirit of TMA, strategy and tactics should be considered at the same time, as much as possible, so that each planning procedure and decision process may benefit from the interaction with each of the others. Tactical issues, however, can be thought of as subservient to strategic issues, in terms of priority, commitment, and timing. They also include more implementable details and are closer to the nuts and bolts of the business.

The time for strategic planning is after answering the following questions:

"What should we build?"
"Why should we build it?"
"When should we start building it?"
"How many should we build per day/month/year?"

At this point, tactical issues must be resolved. A new set of questions then arises:

"How should we design it?"
"Should we make or buy the parts?"
"How should we manufacture it?"
"How should we assemble it?"
"How should we inventory it?"
"What types of machines should we use?"
"How should we arrange our factory?"
"How should we control our manufacturing processes?"

The activities and essential techniques for addressing these and related issues are presented in the remaining sections of this chapter.

A. Design For Manufacturing

An important tactical tool is the philosophy of design for manufacturing (DFM). Although DFM is considered a tactical tool from the perspective of manufacturing planning, it also may be considered a strategic tool (that is, a farsighted or long-range tool) from a product-design viewpoint.

The idea of DFM is an example of TMA at its best: functions that are traditionally considered separate and sequential are now linked into a simultaneous and symbiotic process. In this case, the two functions are product design and process design. Rather than designing a part merely on the basis of its intended function, we design it with an eye to how best to manufacture it.

For example, imagine that several designs have been suggested for a part. Traditionally, the best design is chosen based on such issues as the intended function of the part and the costs of materials. Although these considerations are good and proper, they ignore the manufacturing consequences of the decision. What are the process costs of the selected design? How many machining steps are required? How many assembly steps, testing steps, and packaging steps? How many jigs and fixtures will be needed? How many labor hours? Can the process be automated? Can the design be assembled from standard parts? The trick is to think of the product not as a stand-alone item, but as a part of the entire business enterprise of the corporation.

One approach that improves DFM competency is restructuring of design teams. A product development team should include one or more members from process or plant engineering who will have direct input into all design decisions. At the very least, a process representative should be consulted at regular intervals during the design phase.

This approach is especially important when a major new product line is being introduced and new facilities are being planned. The functions of product design, process design, tool and fixture design, plant layout, vendor selection, and equipment selection can and should be performed more or less simultaneously.

1. DFM Guidelines

What are the rules in designing for manufacturability? Certain guidelines have been established by various industries. Some are more applicable than others in different situations, so they must be adapted to your particular product line and industry. Still, the DFM philosophy is well understood by studying these basic rules.

1. *Reduce the number of parts.* With fewer parts in a product, there are fewer drawings required, fewer assembly steps, and less paperwork. The results are lower error rates in both manufacturing and assembly, quicker production time, and increases in both quality and reliability (because there are fewer parts to fail).

2. *Reduce the number of part numbers.* Not only should each product have as few components as possible, but the entire plant should require as few different types of components as possible. Always use a standard or "preferred" part that is already designed and on hand if it can do the job. This reduces inventory, paperwork, and chances for assembly errors.

3. *Reduce the number of vendors used.* Try to select vendors who can deliver as many components as possible. This simplifies quality validation procedures, reduces the number of shipments needed, and minimizes paperwork and the chances of back orders.

4. *Design for robustness.* Robustness is defined as insensitivity of output to variations in the input. Design the product to function correctly and consistently even if component characteristics vary. (Such variations may be introduced by the manufacturing system.)

5. *Reduce the number of adjustments needed.* Try to eliminate, if possible, the use of adjustable components such as setscrews, tunable electrical components, tensile or compressive fits, and so forth. Attempt to make all assembly steps of the "positive fit" variety, so that as soon as a component is in place, it is ready to function properly. This eliminates the possibility of adjustment errors, increases assembly quality and reliability, and vastly decreases assembly time and cost. This principle is worth following, even if it results in an increase in component cost.

6. *Use foolproof assembly steps.* Use components that fit only one way— the correct way. Do not use force fits. Try to avoid the use of threaded fasteners; snap-fit parts are quicker, cheaper, and easier to automate. Use chamfers and other location features to aid in assembly. Sequence assembly steps so that component B cannot be inserted until component A has been inserted correctly. The fewer the number of possible assembly errors, the fewer will be the number of actual assembly errors.

7. *Design for vertical stack assembly.* Ideally, all components should be assembled along a single axis, like stacking up a sandwich or a pyramid. This reduces the need for reorientation during assembly, makes the process easier to automate, and helps make assembly foolproof (see previous guideline).

8. *Design testing procedures into the product.* Accomplish testing with a minimum of disassembly, reorientation, and measurements. The quicker and cheaper testing is, the more testing you can do.

B. Computer-Integrated Manufacturing

To attempt to discuss manufacturing management without an in-depth consideration of computer applications would be foolish. Computers have so permeated the manufacturing field that many volumes can (and have) been devoted to the various topics included within the general heading of computer-integrated manufacturing (CIM). In this section, we will touch on some highlights and crucial areas and will provide references for further information for the interested reader.

CIM can be arbitrarily broken down into a number of overlapping areas: computer-aided design (CAD), computer-aided manufacturing (CAM), computer-aided process planning (CAPP), computer process control (CPC), and computerized business operations. This section will focus on CAD, CAM, and CAPP, and how they interact with computerized business operations in a general CIM system. CPC topics such as numerical control are discussed in Chapter 5, under "Automation."

1. Computer-Aided Design

Computer-aided design is the application of computer technology to the product design function. In the most trivial sense, CAD replaces the traditional drafting tools of drawing board, vellum, and pencil with a computer-graphics workstation. This innovation alone supplies several advantages:

- Increased efficiency—A drafter or designer can easily double productivity by proficient use of a computer-based drafting system. The investment required to convert a drafting department to computer technology is roughly $20,000 to $100,000 per workstation and one to two weeks of training per operator. This investment rapidly pays for itself, particularly if equipment is used for multiple shifts.
- Automatic documentation—A computer-generated document is easily stored and backed up on a file server system and can be readily distributed to the departments that require it. It will not degrade with time or with multiple generations of copies, nor will it distort when being reproduced.
- Ease of modification—If several similar parts are to be drawn, a single file can be modified easily to represent each version with no duplication of effort.

This list of advantages, however, ignores the fact that a CAD system is far more than merely an automated drafting board. It is a computer-based system, with all the programmability inherent in the power of the computer. The following new capabilities become available:

- Automatic design—Many CAD systems have been developed that automatically draft a component based on a few simple input parameters. For example, a CAD department can easily write a software interface that will query the drafter or designer for the number of poles, coils, and field magnets in a generator, and some basic dimensions, and will then draw it automatically. The time required to draft standard parts using CAD can be reduced to a mere 1 or 2 percent of the time required to draw parts manually.

• Automatic analysis—Similarly, a design that has been created in a CAD system can also be linked to an analysis package to quickly and accurately calculate volume, mass, center of gravity, and so on. Finite-element packages can also be linked to CAD systems to do stress analysis, heat dissipation analysis, and so forth.

• Automatic data base—The CAD system can also be used to perform automatic feature extraction and data base management. Features such as holes, fillets, and finish surfaces can be counted automatically and stored in files to be sent to manufacturing, accounting, or bill-of-materials packages elsewhere in the company.

CAD systems are available in a wide variety of configurations, prices, and capabilities. In general, CAD systems consist of the following components:

1. Graphics hardware.   There must be a high-resolution graphics screen capable of displaying large drawings in fine detail. Many of the best screens provide 1024- by 1024-pixel resolution, with up to 1024 colors to choose from. There must also be a graphics input device such as a mouse, trackball, or digitizing pad, in addition to a keyboard. Specialized CAD workstations are available for tens of thousands of dollars, or simple personal computers (PCs) can be used for smaller jobs.

2. Graphics software.   Specialized workstations come with software provided, or PC-based software can be purchased separately. Either way, the software includes facilities for creating, modifying, and deleting entities such as circles, lines, points, labels, dimensions, and possibly planes and blocks. Many systems allow entities to be grouped into "layers" that can be hidden or revealed independently of the rest of the drawing. The software will also supply some sort of file management tools.

3. File server.   Large industrial systems often supply a separate computer with large amounts of disk space to store drawing files. The workstations then do not need to store files locally, any workstation can access any drawing, and if a workstation is down for repairs, no data is lost or tied up. For PC-based CAD stations, files are generally stored on the individual computers, unless some sort of local area network has been implemented.

4. Graphics output device.   This is usually a large X-Y plotter, capable of reproducing drawings of any size. It consists of a large flat table on which the blank sheet is mounted. A long movable bar slides back and forth across the table in the Y direction, while a pen slides along the bar, supplying the X motion. Usually the pen is selected from a turret of pens, each of a specific color and line thickness. Alternatively, some cheaper CAD systems use dot-matrix printers or laser-writers as output devices, but these tend to be limited in the size of drawings they can output.

2. Computer-Aided Manufacturing

Computer-aided manufacturing is more complex than CAD and covers a broader range of functions. It can be thought of as the link between CAD and CPC: It takes the data base that was generated and stored by the CAD system and formats, transmits, and prepares the information to be sent to the numerically controlled machine tools on the factory floor.

Many incarnations of CAM exist, and each must be tailored for a specific factory and its equipment. Unlike CAD systems, which can be bought off-the-rack by any manufacturing company, a CAM installation is a complex system of many items from many vendors, carefully selected, installed, and orchestrated to implement the automation that the company requires.

CAM is best defined by example. Consider a fairly simple installation for drilling holes in circuits boards. A designer in the CAD department has just completed a drawing of a new circuit board, which consists of a new arrangement of holes and circuits on a standard-sized board. A CAM system for this application might start with a custom program that queries the CAD data base and compiles a list of holes to be drilled. The list includes three numbers for each hole: the X and Y location of the center of the hole from some reference point, and the hole diameter. The list also includes the part number of the standard board from which the part will be produced. This list is transmitted via some network to the shop floor.

An operator, watching a screen, will see that a board has been ordered. He or she selects the required board from inventory and places it in a CNC (computer numerically controlled) drill press. When it is ready, he or she presses a button, and the hole list proceeds to the CNC press, which then drills the required holes. No paper has been generated, and there have been very few opportunities for error. The hole file could even have specified that multiple parts be produced from the same file, which the operator would then have produced.

A more complex CAM system could automate many more steps in the production process and could reach more areas of the factory. Parts could be ordered from inventory automatically and delivered by an automated guided vehicle (AGV) (see Chapter 5) to the appropriate machine tool. Robots could manipulate the parts onto the machines and deliver them to the final assembly area.

As the role of CAM expands in a factory, the overall concept of computer-integrated manufacturing is approached. Other roles that the CAM system can take over on the way to total CIM include computer-aided process planning, cost estimation, part programming, inventory control, MRP (material requirements planning) and JIT (just-in-time) implementation (discussed below), automated testing and inspection, and development of work standards.

3. Computer-Aided Process Planning

Computer-aided process planning is one of the most vital links between the design and manufacturing functions in a CIM factory. By definition, process planning is the sequencing of the manufacturing steps to convert raw materials into products in accordance with the design specifications. It is often thought of as something of an art form and can be very time-consuming. For this reason, and because of its connection with many factory functions, CAPP is an important and far-reaching concept.

CAPP, and process planning in general, addresses the following questions:

1. What raw materials should we start with?
2. What machines should be used?
3. In what order should the parts be machined?
4. In what order should the components be assembled?

In a noncomputerized process planning system, these questions are answered by manufacturing personnel and are documented on a route sheet. This sheet is duplicated and distributed to all stations involved, resulting in a large paper trail.

CAPP improves the process in two ways: it reduces the paperwork, and it removes the human element from the planning process. The first task is a simple matter of information technology and data distribution, but the second is much more difficult. Development of the process plan is a task of synthesis; various options are weighed and an overall plan is created that will result in the finished product. It is a process that requires intelligence and, as such, cannot be automated without a great deal of effort. There are two common approaches to developing a CAPP system that can generate a process plan: the derivative approach and the generative approach.

A derivative CAPP system, also called a retrieval system or a variant system, seeks to create a new process plan out of an old one. It takes as input the CAD drawing of the new part and analyzes the part in terms of general shape, types of features, and complexity. This analysis results in a group technology (GT) code number that describes the part in a general, feature-oriented manner. The CAPP system then searches the company's part data base for a previous part with a similar GT code. When an old part is found with a sufficiently similar code, the route sheet for that part is retrieved and modified into a route sheet for the new part. The new part and its route sheet are then added to the data base for future use. If no closely matching GT code can be found, the CAPP system either signals for an operator to develop the process plan by hand, or defaults to a generative mode.

A generative CAPP system takes a "from the ground up" approach, generating a process plan completely from scratch. This system is an example of an artificial intelligence program and is usually created using expert system programming techniques (see Chapter 4). Basically, the program contains "rules" that relate product features to process steps. For example, one rule may state that if the overall shape of the part is cylindrical, the process plan should start with round bar stock, cut to the proper length. If it is cylindrical with varying diameters, a lathe step will be included. If it is cylindrical with holes, a drill press will be used, and so on.

Regardless of the type of CAPP system used, a route sheet is prepared, stored, and routed to the necessary departments. It can be printed out and sent to each recipient on paper in a less integrated factory or sent electronically in a more fully CIM-oriented factory.

The use of a CAPP system has other advantages besides reducing paperwork and saving time. One is standardization. Two human process planners might come up with two different process plans for the same part, but a CAPP system will always provide the same plan. This can standardize factory operations and reduce time wasted in refixturing and retooling. Further, two similar parts will generally have similar process plans, which might not be the case with manual process planners. Finally, since CAPP necessarily runs on a computer, an interface is automatically provided to other parts of the CIM operation of the factory. The CAPP system is easily interfaced to an MRP system, inventory system, customer billing system, and so forth.

4. CIM in General

This subsection has discussed some of the elements that are integrated into a CIM system. Although they have been defined separately, it should be remembered that none of the elements reaches its full potential unless it is integrated into a powerful, connected system.

The remaining elements to be discussed are the dissemination and use of the information developed by the CAD, CAM, and CAPP elements by the rest of the factory. Two topics in this area are discussed: MRP and JIT control. However, before diving into discussions of these high-level concepts, there is one more element of the CIM factory that must be covered: the machines that will do the actual production work. These are discussed in the next section.

5. Flexible Manufacturing Cells and Systems

Flexible manufacturing is the concept of using capital equipment such as machine tools, robots, and controllers to produce a variety of products whose mix is not known in advance. In a factory that produces large volumes of a small number of parts, there is no need for flexible manufacturing. A factory that produces a variety of parts, but in well-defined and preplanned proportions, has no need to adapt to changing demand. These types of factories are still well advised to make use of TMA techniques, but need not worry about the material in this section.

However, many industries do not have the luxury of advance warning of their production requirements. In the quest for ever-decreasing lot sizes, set-up times, and inventories, flexibility of production capability is becoming increasingly attractive to more and more companies.

What is this flexibility, and how is it acheived? The flexibility we are discussing involves the ability to produce

1. A variety of products
2. A variety of product mixes
3. New parts with minimal lead time

This implies that we do not know, in detail, what we will be called upon to produce, or how much, or when. Our production equipment must be flexible enough to perform whatever tasks are necessary, within reason, to meet the demands.

This flexibility is achieved by the use of a flexible manufacturing system, which is a group of equipment, generally under computer control, that can produce a variety of products with a small setup time between products. Ideally, the system should be able to produce one of each product at a time, and produce it profitably.

A term often used to describe this type of system is *flexible manufacturing cell* (FMC). The difference between a flexible manufacturing cell and a flexible manufacturing system is largely arbitrary, but in general a system is thought of as a larger and more flexible installation than a cell. A flexible manufacturing cell usually contains a small number of machine tools (say, a lathe, a mill, a band saw, and a robot) under the direct control of one computer and designed to produce one part "family" of very similar parts. The term *flexible manufacturing system* (FMS) generally designates a larger collection of machine tools, along with material handlers, some sort of storage and retrieval system, several controllers, and a central supervisory computer, all geared toward the production of several different part families.

Possible components of an FMS or an FMC include all the equipment used in any type of manufacturing. However, the most useful items to be included are the multifunction and programmable versions of these devices, such as CNC milling machines, machining centers, and turret lathes, along with the appropriate tool-changing devices. Because FMSs are often run with minimal supervision, automated inspection equipment—such as automated coordinate measuring machines—are often included as well.

The functions of an FMS need not be limited to machining. They may also include tooling for sheet-metal forming, injection molding, forging, and joining processes. Larger systems require an extensive amount of material-handling equipment, such as conveyors and AGVs for moving material from one machine to another and robots for loading and unloading parts to and from each machine.

Another possible component of an FMS is one or more human operators. While FMS usually implies a high level of automation, that is not a requirement. A skilled human operator who is familiar with the process plans of all types of parts to be produced can replace the material-handling equipment and greatly reduce the cost of the FMS.

How is an FMS designed? There is an optimal selection of machine tools and material handlers for any one specific application, based on the products to be produced and a best guess of the overall product mix required. Upon determining that an FMS is the appropriate approach, it is necessary to select the equipment, arrange it in the most efficient manner, and be able to control it for maximum throughput and machine utilization.

The first step in FMS design, therefore, is to identify the requirements for the system as specifically as possible. Since the system, by definition, will be flexible, these requirements will naturally be ranges and approximations rather than exact numbers. The questions to ask are:

1. *How many different products will be produced and how different will they be?* For our purpose, these are really the same question; they seek to determine the amount of flexibility needed in the FMS. The general rule is that efficiency decreases with increasing flexibility. When the need for flexibility becomes too large, the loss of efficiency makes the FMS approach infeasible. On the other hand, if very little flexibility is needed, a dedicated manufacturing setup will be more efficient than an FMS could ever be.

The moral of this trade-off is that the FMS implementation is most attractive when a moderate amount of flexibility is required. If only one or two very similar parts are to be produced, the FMS approach should be abandoned in favor of a less flexible but more efficient installation. If there are a large number of widely varying products, a sufficiently flexible single system will end up with poorly utilized equipment. In this case, the operations would be best broken up into separate cells, some flexible, some dedicated, as circumstances dictate.

Well then, what amount of flexibility is appropriate for FMS operation? How many products should there be, and how different should they be, to require sufficient but not excessive flexibility? Unfortunately, there is no general rule, since the actual numbers depend upon the nature of the industry in question.

2. *What volume of production is required?* This must be estimated from the product mix we intend to produce, the market conditions that we expect to prevail, and the level of production that we wish to attain. The production volume does not have a particularly strong influence on whether or not to

use FMS, but it must be known to determine capacities and amounts of equipment to purchase.

3. *What is the anticipated product mix?* Of the different products the FMS will produce, approximately what percentage of the production volume will be devoted to each product? This question cannot be answered with precision. If it could, we would have no need for flexibility at all; we want the flexibility so that we can respond to changes in demand for each product. But we must decide what range of variation is likely for each product, so that we will be prepared to handle it.

Once these questions are addressed to the best of our knowledge, detailed specifications for the cell or system can begin. There is no general rule for FMS design, as distinct from design of any other production system, with regard to selection and layout of the equipment. The best approach is to use standard industrial engineering techniques, as presented in Chapter 5, to develop several candidate system designs. These various alternatives should then be evaluated using simulation techniques (also discussed in Chapter 5) to determine which best meets the requirements for throughput, machine utilization, efficiency, and adaptability to the range of product mixes that are anticipated.

Once the FMS has been designed, specified, and built, it must be controlled. This is generally accomplished by one or more computers, operating at several levels of supervision. In the case of an FMC, with a small number of machines in one localized area, a single dedicated computer of midrange capability is usually sufficient to handle all control tasks. A larger FMS, composed of several cells and an extensive material-handling system will require an additional level of computer supervision and will usually need several autonomous controlling computers with one supervisory mainframe to keep track of them all.

Whatever the arrangement of computer architecture, there are several levels of control function that must be met.

1. Shop floor control. This level is necessary only for a large FMS and is responsible for keeping track of the activities at each individual cell and location within the system. It is usually performed by a mainframe computer with connections to each of the individual cell controller computers. It tells each cell what to produce, when, and how many. It controls the AGVs and conveyors and keeps track of the parts and raw materials in the automated storage and retrieval systems. It uses an optimization procedure to allocate tasks for maximum throughput of the system as a whole. It also keeps track of the performance of the individual cells and may include some artificial intelligence programs for adapting to changing conditions and improving performance. It will probably be connected to the company's financial and business records, as well as to the CAD system.

2. Cell control. In the case of an FMC, this is the top level of control. In the larger scale FMS, this level is directly below the shop floor control. It is responsible for the operation of an individual cell or unit in the FMS, such as a storage and retrieval system or a cell of machines dedicated to a particular family of related parts. It is usually implemented with CNC technology, a mini- or supermicrocomputer that can control several machines at once, in real time. It stores the process plans for each part made at the cell, as well as instructions for automated inspection if the cell has that capability. It also keeps statistics on the performance of the machines within the cell and maintains control charts based on the results of the automated inspection.

3. Machine control. Each machine within a cell must be individually controlled. In a less automated system, these would be the functions carried out by a programmable logic controller (PLC) or NC unit. In an FMS, these devices may still be used, under the supervision of the cell control computer. Alternatively, a PC or other microcomputer may be used to link each machine to the cell controller, or the cell controller may be linked to each machine directly. Either way, it is a separate function, with the tasks of running the part programs and monitoring the machine performance, detecting tool wear and breakage, and handling any error conditions that might arise.

C. Material Requirements Planning

At this point, our factory is in pretty good shape. We have designed our parts for maximum manufacturing efficiency. We have used CAD to generate part files, CAM to send the part files to the factory, and CAPP to decide how best to produce and assemble the products. We have specified, purchased, arranged, and connected our manufacturing equipment in the most efficient manner. We are only one step away from running a neat, tidy, profitable, and efficient factory: we have to tell every person and every machine what to do! We have to fire the starter's gun and make sure that all our carefully planned activities occur at the correct times, in the correct order.

Material requirements planning, or MRP, is one part of this execution and control function, and a very important part. MRP ensures that we have the materials we need and that we have them on time. It sounds simple, and in principle it is. However, in a large and complex factory, with dozens or more products, each composed of hundreds or more components, this becomes a gargantuan task. Fortunately, it is all based on simple math and bookkeeping, and so can be explained, if not implemented, with a minimum of difficulty.

An MRP system is based on one simple equation:

[what we need] = [what we want] − [what we have]

In other words, we must calculate how many units of raw material and components are necessary to manufacture our planned production volume, and subtract from that the number of units that we have, or will have, in inventory. This leaves us with the number of units that must be acquired. It sounds simple, but there are complicating circumstances that must be considered. One is magnitude: we may have a large number of products, composed of a large number of components. Another is timing: we must order components early enough for them to arrive on time, and we must order raw material early enough to produce the components we need on time. Finally, there is time phasing: we do not need an entire year's worth of material all at once, but will want to spread it out on a period-by-period basis. We will examine each of these items individually.

1. Master Production Schedule

Our MRP system is driven by a master production schedule, which is an output from our strategic manufacturing planning functions. Someone has decided how many units of each of our products we must have, ready to deliver, for each period of time. The methods employed to develop this master production schedule were discussed earlier in this chapter and come from sales and demand forecasts, firm commitments from customers, desired inventory levels, and so forth.

A master production schedule is a two-dimensional matrix: it expresses the quantity desired for each product during each time period. The size of a time period varies from one company and industry to another, but is typically a week, month, or quarter. Daily or yearly time periods are possible, but are rarely practical. In the case of a monthly period, the master production schedule would dictate the required production of each of our products for each month. These production periods are also known as "time buckets."

This raises the question, how many time periods or buckets should we include in our schedule? This also varies from company to company, but should be as large as each company finds feasible. The schedule can always be updated if demands should change. The length of the master production schedule is called the *planning horizon.*

To summarize, our master production schedule is a two-dimensional matrix, with each product we sell listed down the left side and each time period within our planning horizon listed across the top (see Figure 4). These numbers are used to derive the "what we want" values in the MRP equation given above.

2. Bill of Materials File

Another input to our MRP system is called the bill of materials (BOM) file. This is used to calculate the details of the "what we want" values in our MRP

PERIOD	JAN	FEB	MAR	APR	MAY
X-11 Computer	300	400	500	500	500
X-15 Computer	800	800	800	1000	1000
X-22 Computer	400	200	200	400	500

Figure 4   Master Production Schedule

equation. While the master production schedule tells us what we want to sell, the bill of materials tells us what we want to produce it from.

There is a separate BOM file for each product in our catalog. In a well-automated factory, a BOM file can be produced automatically from the CAD data base for each product. Each file contains a hierarchical list of each subassembly in the product, each subsubassembly in the subassembly, and so on, down to each individual component. A typical BOM format is called the *indented bill of material* list, as shown in Figure 5.

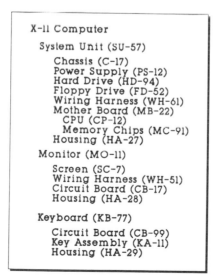

```
X-11 Computer

    System Unit (SU-57)

        Chassis (C-17)
        Power Supply (PS-12)
        Hard Drive (HD-94)
        Floppy Drive (FD-52)
        Wiring Harness (WH-61)
        Mother Board (MB-22)
          CPU (CP-12)
            Memory Chips (MC-91)
        Housing (HA-27)

    Monitor (MO-11)

        Screen (SC-7)
        Wiring Harness (WH-51)
        Circuit Board (CB-17)
        Housing (HA-28)

    Keyboard (KB-77)

        Circuit Board (CB-99)
        Key Assembly (KA-11)
        Housing (HA-29)
```

Figure 5   Indented Bill-of-Material File

The level of automation and data integration in a factory greatly influences the ease with which the BOM is integrated into the MRP system. In a completely manual system, large amounts of bookkeeping must be performed by hand, by armies of clerks, data entry personnel, and bookkeeping personnel. As a complete CIM system is approached, more and more steps in the information-processing function can be automated, resulting in a hands-off computer-driven MRP system.

3. Inventory Record File

The other input needed for the MRP equation is the "what we have" term. In other words, which of the components and raw materials are already in inventory? More precisely, how many of each item will be on hand, and not already spoken for, at the time period when it will be needed? Time phasing must be followed scrupulously at this stage of the MRP calculations.

This information is stored and reported in the inventory record file. This file maintains the information of how many of each item of raw material, component, subassembly, and final product will be on hand for each time period throughout the planning horizon. The format varies from one installation to another, but there are generally lines for total inventory, projected inventory, inventory already committed, and inventory available for use in future production.

Again, the inventory records can be completely manual, highly computerized, or anywhere in between. The MRP system can function with any type of inventory system, but is most easily performed with a computerized inventory system.

4. Timing

We now have the information we need about what we want and what we already have. We are almost ready to calculate what we need to acquire. The only other concern is when we need to have the parts on hand.

A fail-safe answer to this question would be: get it now! If we order all parts that we expect to need as soon as we realize we will need them, we will rarely be caught short. True enough, but then we are faced with unnecessary carrying costs, not to mention vastly increased storage space needs. This approach also encourages sloppy production planning techniques, as we shall see in the next section on JIT planning. We will be in better shape if we plan to receive our required materials at the beginning of the time period in which they will be needed.

The final piece of information we now need is lead times. When we place an order for a quantity of raw material or components, there is some time lag before we take delivery. This is called *ordering lead time*. We must order our required materials at a time that is earlier, by an amount of time equal to the ordering lead time, than the time at which we must have it.

Similarly, when we have all the materials needed to manufacture a product, it is not instantly machined, fabricated, assembled, tested, and packaged. All these steps take time, and the time will vary from one product to another. This time is known as *manufacturing lead time* and is calculated from the process plan. Therefore, the time at which we must have the materials is earlier, by an amount of time equal to our manufacturing lead time, than the time at which we must deliver it.

Once all these items of information have been considered, we will be able to calculate exactly when to place the orders for each of the materials we want delivered to our factory. If nothing goes wrong, each item will arrive precisely in time to allow us to meet our master production schedule. The overall relationships between the various segments of the MRP system are shown in Figure 6.

Of course, it is not quite that simple. The process described above takes the master production schedule and "explodes" it, in accordance with the

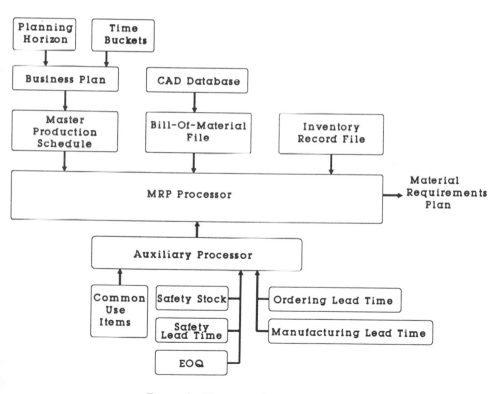

Figure 6 Elements of an MRP System

BOM, into a huge list of raw materials. It also spreads the schedule out over time to accommodate the various lead times involved in ordering and manufacturing each component and subassembly. The result is a highly scattered and fragmented list of items and order times. A certain amount of consolidation is often in order.

One concept to consider is *common use items*. As mentioned under the "Design for Manufacturing" subsection, we have attempted to design our products with common or "preferred" parts whenever possible. If we have, our MRP file may specify many separate orders for the same item. Obviously, these should be consolidated into one order if they all occur within the same period.

Another consideration comes from standard inventory theory concepts such as economic order quantity (EOQ). If our MRP system tells us that we need to order 9,000 units of some item in a given month, but we get a volume discount for ordering 10,000, it may be advantageous to stock up, despite the slightly increased carrying cost.

A further issue to consider is uncertainty. Is manufacturing lead time really a constant? How about ordering lead time? How crucial is each part to the overall process plan? When there is some doubt about the time required to finish a production plan, it is sometimes advisable to use *safety lead times*, a cushion of time to make up for possible slippages in the schedule. A similar concept is *safety stock*: excess inventory to account for uncertainty in demand, potential quality problems, or uncertain yield.

5. MRP-II

One other concept worthy of note here is the newer version of MRP known as MRP-II, or manufacturing resource planning. This is basically an expanded version of material requirements planning, with extra functions integrated into the planning procedure to convert it from a tactical tool to a strategic one. To convert an MRP system to an MRP-II system, it is generally linked with the financial plans of the company and often is interfaced to a computer simulation, as discussed in Chapter 5. This is also known as *closed-loop MRP*. Its overall functions and structure are not as well defined as the standard MRP concept and are beyond the scope of this discussion.

D. Just-in-Time

There is much talk today about the just-in-time approach to manufacturing planning and control. It is often hailed as revolutionary and as the only competitive way to run a factory in the modern economic environment. Few people, though, really understand what a JIT system means or what it includes.

While a JIT system can revolutionize the operation of a factory, it is not based on any new concepts. JIT must be implemented with all the traditional techniques that most production people have known for years. What it changes is the philosophy that is used to run the factory, not the tools. The tools are used with a new attitude, which enables them to be used much more efficiently.

The best way to understand how JIT works is to imagine learning how to swim. One way is to take lessons in a pool, getting instructions from an expert teacher, practicing each stroke and position, gradually learning the components, and honing your skills until you are ready to attempt a swim in water that is over your head. Another way is to go out in a boat to the middle of a lake and jump overboard: you will learn to swim, or you will drown.

JIT is like jumping overboard. It does not include any new skills that you did not already know, nor does it require any. Rather, it puts you in a position where you must use what you know, and in the most efficient possible manner, or disaster will befall. This sounds dangerous. It is! Many would-be JIT practitioners have fallen on their corporate faces before getting their JIT system working properly, but when everything comes together, they have an efficient, high-quality manufacturing process.

The trick, then, is to throw yourself into the manufacturing lake and put yourself into a position where the necessary skills and techniques will come into use to save your economic life.

1. The Lake

The "lake," or the situation that forces efficient operations in a JIT system, is a lack of inventory. The JIT philosophy sees inventory as an unnecessary evil: it artificially buoys up an operation that would otherwise sink and hides the problems that would otherwise be apparent. All efforts, therefore, are focused toward the reduction of inventory, in terms of both finished products and work in process.

This reduction has immediate benefits in the reduction of carrying cost and storage space and in an increase in responsiveness. But these are minor benefits compared to the other changes that will result. The attempt to survive without the crutch of excess inventory improves quality, reliability, throughput, and profits.

How is inventory a crutch? It allows mistakes to go unnoticed. Suppose we are to deliver 100 units of a product at a specific time. If we have only 100 units, they must all be good. If 2 percent are defective, we will be two units short in our delivery. We have drowned instead of swimming. But we have detected the defectives, can trace them back to their cause, and can ultimately cure the problem. If we had maintained an extra 100 units in inventory, our shipping department would have "borrowed" the extra two it needed. The defects would not have disrupted our operation, but we also would not be alerted to the need for improvement.

This principle also holds for work-in-process inventory at each step of the manufacturing process. Each department consumes a little more material, and uses a little more labor, than should be necessary. Finding all these little inefficiencies is virtually impossible, because they are spread out over the entire operation. No single problem is large enough to stick out, but the cumulative effect is that we are carrying a large amount of waste in the system, in terms of materials, effort, and time, without ever knowing it is there.

2. The Toss

How do we toss ourselves into the lake? The key is a switch from a "push"-based system to a "pull"-based system. Each step in the manufacturing process must be based on a signal from a downstream department, signifying a need for the upstream department to perform its function. This system is based upon demand, rather than supply. When department $N + 1$ requires 25 units of work from department N, department N will produce it. If each part requires two components, department N will "pull" 50 units of components from department $N - 1$ (see Figure 7).

How is this an improvement? For one thing, no station is doing more than is necessary. Department N is not allowed to produce a 26th unit, so it does not. It cannot make a few extra "just in case" there is a problem. (Just-in-time is often called the antidote to the just-in-case philosophy.)

Another forced improvement is quality. If department N must deliver 25 units, and is not allowed to produce a 26th unit, all 25 had better be good! Quality has been forced. This does not mean that 100 percent quality will occur automatically; it just means that if it does not, the repercussions will be felt all the way down the line to shipping, billing, and accounts receivable. The problem will be noticed, and steps will be taken! JIT does not specify what those steps should be; that is up to the individual process engineers and will generally be standard techniques that have been known for years.

3. The Implementation

A JIT system is implemented by putting this demand-based system into operation. This requires redesign of many of the process planning procedures. All the changes detailed below lead to a decrease in the inventory level of finished goods and work in process and to an increase in overall quality. The following techniques must be implemented:

a. Reduction of Lot Size.   The ideal lot size for a JIT system is one. Sometimes this is just not feasible, but often it is. In any event, the lot size should be reduced as much as possible, and policies should be instituted to constantly attempt to reduce it further, until a lot size of one is reached.

Under no circumstances should a lot size exceed 10 percent of a day's production. This 10 percent value is a good starting point for a new JIT system,

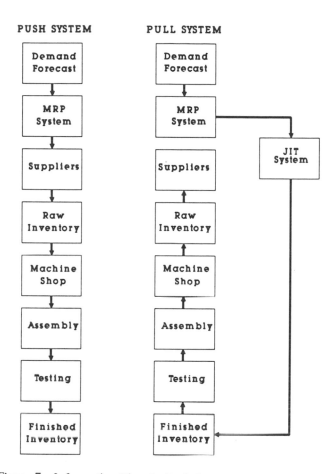

Figure 7   Information Flow in Push System vs. Pull System

and it should be constantly lowered. What if a department produces less than 10 items per day? Is a lot size of less than one feasible? Certainly, since any large, complex unit will be composed of subassemblies or produced by a number of steps. These should be broken down and implemented as separate lots.

b. Elimination of Discrete Batches.   Production should be thought of as a continuous flow, rather than as batches of a discrete number of units. This is literally true when lot sizes have been reduced to one and is approximated when lots are small enough to be completed in less than an hour. The smooth flow of products allows any level of demand to be met without waste and allows smaller and smaller orders to be produced profitably. This also helps customized orders to become more feasible.

c. Production of Mixed Products Simultaneously. It should be possible to produce all products at one time, in any possible ratio of individual levels. Again, this is strictly possible only when lot sizes have been reduced to one, but is reasonably approachable when the 10 percent rule has been followed. A mixed mode of production also allows orders of any size or specialty to be met profitably, with minimal waste of effort or production.

d. Reduction of Setup Times. This is the key to many of the above steps. One of the main things that causes large lot sizes is the overhead of setting up equipment for a specific production process. If setup time is zero, profitable lot size, by definition, is one. As long as setup time is sufficiently small, though, sufficiently small batches will result.

Often, setup time leads to a trade-off in the production engineering phase of process design: Is it worth investing in more expensive equipment to reduce the setup time of a specific item? If lot sizes are to be large, the more expensive flexible equipment is rarely justified. If lot sizes must be small, it generally is not worthwhile to spend more for specialized equipment such as multipurpose jigs and fixtures and programmable machinery. In a JIT system, our primary goal is to reduce lot size. Therefore: get the equipment. The payoff will come in the benefits to be reaped farther down the road.

e. Commitment to "Zero Defects" Goal. Again, we have a situation where a common trade-off must be resolved in light of our new philosophy. Suppose we have two options on our purchase of production equipment: one system of machinery is fairly inexpensive, but will produce product that is defective 2 percent of the time. The other brand of machinery costs twice as much, but will yield good product 99.99 percent of the time. Is the better equipment worthwhile? Suppose the cheaper machine costs us $5 per day in wasted materials and energy, but is cheaper by $100,000. In a shortsighted sense, the waste is justified. But the JIT philosophy says it is not. Allowing the waste into the system would destroy the demand-based "pull" system, would create excess work in process, would obscure inefficiencies in the system, and would undermine the entire procedure.

Quality improvement puts our system into an upward spiral of further improvements. As quality improves, our batch sizes decrease, our inventory decreases, subtler problems become apparent, and our control over the process increases. True JIT is more closely approached, and further quality improvements follow.

f. Increased Worker Understanding. Workers in the factory must understand more of the production process than just one workstation or department. There are two reasons for this. First, personnel are working on a pull system: each station will function only when the next station downstream requests a batch. When this job is done, what will the workers do? In an ideal situation, they move to any other station that has work to be done. It is the workstations and their specific functions that must wait for a pull, not the employees. The more tasks an employee is able to perform, the more fully utilized that person will be. He or she will also see the overall operation of the factory that much better.

The second reason is based in the JIT philosophy of continually striving for improvement. The best source of ideas for process improvement are the employees who are doing the work. The more stations an employee can operate, and the broader his or her view is of the operations at large, the more useful his or her insights and suggestions will be. One of the main reasons for reducing inventory was to make problems more evident. The other side of the coin is to enable the workers to see the problems when they surface.

g. *Commitment by Management.* The importance of this element cannot be overstressed. JIT cannot be implemented in one department at a time; it must follow the production process of at least one product line from beginning to end. This will require a fairly high level of management agreement right from the start.

The first thing that will happen when a product line is converted to JIT control is that it will fall flat on its face. Production schedules will not be met, deadlines will be blown, deliveries will be late, and customers will be dissatisfied. This cannot be helped! It is the only way that problems and inefficiencies will be found! It is the essence of the JIT process.

The second thing that will happen is that management will want to pull the plug on the whole idea. They will see that the state of the company is going from reasonably good to horrible, and they will want to give up. This must not be allowed. Management must be made to understand, right from the start, that this is an expected, indeed necessary, phase of the JIT conversion.

The third thing that will happen is that the process will improve again. As problems are found and corrected, production will begin to catch up with the required schedule. Process efficiency and product quality will vastly exceed the pre-JIT levels. Management will see that its commitment was justified.

4. The Pull

How is this pull system instituted? Two things must be done. The master production schedule must be adapted to put the initial pull into the production system, and each department or workstation must be given a way to pull on the department upstream from itself.

The master production schedule was discussed in the previous subsection on material requirements planning. This schedule is used to give our production system the initial tug to get things started. The production schedule is generally written in terms of numbers of each product to be delivered each week, month, quarter, or some other convenient period of time. These demands must be subdivided to daily production requirements according to some scheme. For example, a monthly schedule may specify that 1,000 units of product A be produced in a given month. If that month has 20 work days, we could merely say that production each day should be 50 units. We could then do the same with the other products in the schedule.

On the other hand, we could take the opportunity to add some intelligence to our production schedule. If item A is a hotter seller than our other products, we may wish to produce the entire month's worth as soon as possible, then spread out items B through Z evenly throughout the rest of the month. We may wish to keep production of certain products in step with each other, or to delay a specific product to the end of the month. Whatever product blend we wish to require on each day, our "continuous production mode" allows us to do it.

Whatever method we choose to employ, we will generate a production requirement for each production day on the calendar. This requirement is sent to the last station in the production process for each product. The final station might be inspection, assembly, packaging, or whatever our process plan has specified. JIT does not require us to change the process plan, only how it is followed. Also, remember that there is no allowance for safety stock or safety lead time; we are in a no-defects, zero-inventory mode, and we must stick with it.

Now we have injected a one-day pull into the production system. We will do the same each day, as planned when we divided up the monthly (or whatever time period) production plan. Of course, we are free to alter our daily schedule at will, should our demand forecasts be updated, or if previously unanticipated orders should be received. That is another JIT advantage.

Once this one-day pull is started, how is it propagated through the entire production process? There is no requirement that this be accomplished in a specific way, and each company is free to establish whatever procedure works best for it. Some of the more popular are summarized here.

a. The Verbal Method. In a small facility, there may be no need for formal procedures. Merely walking over to the upstream department or picking up the phone may be sufficient to get the message across. This is fine for small job shops and low-volume producers, but is not recommended for larger operations.

b. The Floor Outline Method. A slightly less casual method is to draw an outline of the floor between each two successive departments or workstations.

This outline should enclose an area just large enough to hold one lot's worth of work in process. The downstream department draws its input from within the outline. The upstream department keeps an eye on the area. When the outlined area is empty, that is the signal to go to work filling it up again. When sufficient output has been produced to fill the outline, the downstream department stops.

The downstream department must understand that it is not to start work again until the area is empty. Otherwise, there will be a permanent inventory of work-in-process, and the JIT philosophy will have been violated.

c. The Empty Pallet Method. A variation on the floor outline method is to use an empty pallet or container. Exactly one container exists for each interdepartmental relationship. That is, there must be one container to shuttle between departments $N - 1$ and N, one to shuttle between departments N and $N + 1$, and so on.

The containers are sized to hold exactly one lot of the work-in-process that travels between the two departments. Department N draws its input materials from the container until it is empty. It then sends it back upstream to department $N - 1$. The empty container is the signal for department $N - 1$ to begin production again and fill it up. When full, it is sent downstream to department N.

This is basically the same as the floor outline system, except that it can be used for departments that are not near each other. It is therefore more convenient for large factories with complex process plans.

d. The Kanban Method. This is an even more sophisticated technique, but is still based on the same principles as the methods presented above. Instead of sending empty pallets between departments, authorization cards (*kanban* in Japanese) are sent. When a downstream department is ready for one lot of input, it sends a kanban to the previous department upstream, authorizing one batch of work.

The kanban system is more flexible than the previous methods because it is based purely on exchanges of information to signal production runs, not physical entities such as pallets. The kanbans can even be electronic signals sent over a computer network. This allows extremely large factories with remote and highly interconnected production processes to operate in the JIT mode.

Some systems use a dual kanban approach. One type of kanban is a production signal, telling the upstream department to begin production. The other type is a transportation kanban, authorizing the movement of either an empty container to an upstream department or a full container to a downstream department.

5. Results

When a JIT system is begun, the initial result will be chaos. As stated earlier, the factory is not able to handle the new requirements and so falls flat on its face, as surely as if a crutch had been removed. That is exactly what has happened, as the crutch of inventory has been eliminated.

The final result, however, will be wholesale improvements in efficiency of the overall operation. These include:

1. Vastly improved quality. Since there is no room for defects in a JIT system, they are necessarily eliminated or vastly reduced.

2. Minimized inventory and storage costs. Since inventory of finished products and work in process are virtually eliminated, no money is tied up in unsold material.

3. Increased equipment utilization. Since setup times must be reduced to operate in JIT mode, the equipment spends more of its time in useful production work.

4. Minimized waste. Since there is little or no room for error in the new system, losses due to wasted material, energy, and time are eliminated as much as possible.

5. Reduction of paperwork. Since there is no longer any significant inventory, there is no need to keep track of it. There is also no need for a central production planning facility to tell each department what it should be doing, since the pull system distributes that information as it is needed.

6. Constant Striving

Finally, remember that most JIT systems are never completely implemented: there is always room for improvement. The JIT philosophy is asymptotic. It specifies zero defects and lot sizes of one. If these are somehow attained, great! Generally, they will only be approached. That leaves constant room for improvement, which *must* be attempted. Even if further reduction in lot size and error level cannot be realized, the very act of trying improves the efficiency of the process. This is very important: *trying* to improve those two factors *will* improve throughput, utilization, and overall efficiency. This sounds like cheerleading, but it is not. It is the essence of the JIT philosophy.

E. Economic Analysis

All tactical manufacturing planning decisions have economic consequences that must be considered. Naturally, decisions are slanted toward those alternatives that maximize profits. An understanding of basic economic principles is therefore essential to sound tactical planning.

The fundamental arithmetic of economic planning is known as *engineering economics* and focuses on the time value of money. Many excellent texts are devoted to this topic, as well as any good course in microeconomics, so we will not belabor it here. A few vital points, however, are addressed.

1. Time Value of Money

The foundation of engineering economics is the idea that money has a time value: the longer you have it, the more it is worth. If you received $100 today, that would be better than receiving $100 a year from now, even if there were no inflation.

Why is this? It is because money is a piece of production equipment. It can be used to produce products that can be sold for a profit. If we had a million dollars, we could buy a factory, operate it for a year, pocket the profits, and then sell the factory. If we sold it for exactly a million dollars (no inflation and no depreciation), we would be back where we started, but with the money generated as profits in our pocket. We used the million dollars, indirectly, to generate more money.

If we had kept the factory for two years, we would have made more profits. If we had kept it only for six months, we would have made less profits. We can see that the longer we have use of the money (or the equipment we can purchase with it), the more profits we will reap. These profits represent the time value of that original million dollars.

Where did we get the million dollars? Most likely, we borrowed it from a bank. The bank did not loan it to us for nothing; that would have been unwise. It could have used the money to buy a factory itself. So to make it worth the bank's while to loan it to us, it requires that we pay them. The longer we keep the million dollars, the more they want us to pay. In essence, we are renting the million dollars, just as we could have rented any other piece of production equipment. When you rent money, the rent is called interest.

The calculations involving interest rate are called *compound interest formulas.* They are not complex, but are beyond the scope of this discussion; the reader who is unfamiliar with them is referred to the texts listed at the end of the chapter.

2. Selection of Alternatives

The task of engineering economics is to evaluate various production alternatives and determine which one gives the most advantageous time value for the money involved. The issues commonly considered are

1. Initial Cost.   How much money is required up front to implement a specific alternative or plan?
2. Interest Rate.   How much "rent" will we be paying on the money that we will use to implement the plan?
3. Annual Cost or Benefit.   What will the monetary consequences of the plan be on a yearly basis? That is, by how much will it increase our profits? Or, alternatively, how much will it cost us each year?

4. Economic Life.   How long will this plan or investment last? How long can we expect the profits to keep coming? When will we have to reinvest to keep our enterprise going?
5. Salvage Value.   After the economic life of the investment has passed, will we be able to sell the used equipment for any significant amount?
6. After-Tax Cash Flow.   How does the investment affect our income tax status? Is the investment depreciable? Is this the best time to implement it?

Each of these values must be determined as closely as possible for a reasonable economic analysis to be made of a candidate plan. The results of the analysis will tell if the plan is profitable and, if several alternatives are available, which is the most profitable.

In terms of manufacturing planning, each of the factors listed above can be complex and can involve many subproblems. Some of these subproblems are discussed below.

a. Initial Cost.   This value can usually be determined fairly straightforwardly. If we are considering the purchase of a piece of production equipment—for example, a lathe—several vendors will probably offer acceptable items. The prices will be well defined, but may be negotiable.

If a vendor will give a trade-in allowance on old equipment that we are replacing, this amount may be deducted from the cost of the new equipment in the economic analysis. Since trade-in values may vary from one vendor to another, it is important to determine the amount separately for each alternative.

It is also important to consider the start-up costs associated with installing and debugging the new equipment, as well as costs for retraining employees and the production time lost during the changeover.

b. Interest Rate.   The interest rate to be paid on an investment can also be determined simply, but first we must determine the actual source of the capital funding to be used. In the case of a bank loan, the interest rate will be stated explicitly. However, many capital expenses will be paid for out of a company's cash reserves. In this case, the cost of the funds is the loss of the interest the cash is currently earning. Obviously, it would be unwise to remove funds from an investment portfolio that is yielding a 10 percent return, just to invest it in equipment that will return 5 percent. This would be justified only when the equipment is absolutely necessary to stay in business and to protect or support other, more profitable investments.

Another factor to consider is variability in interest rates. When an adjustable rate loan is used to finance new equipment, a best estimate must be made of the interest prevailing in each future year. These varying values must then be used in the economic analysis.

c. Annual Cost or Benefit. The impact of an investment on annual cash flow can be difficult to determine. A new piece of machinery will affect overhead costs in terms of energy consumption, raw material use, insurance, maintenance, floor space, and work-in-process. It may also affect labor requirements, either positively or negatively. The effect on overall factory efficiency will be difficult to determine, unless the new equipment is replacing an older machine that was performing precisely the same function.

Other effects on annual cash flow can be even more difficult to determine. These include such factors as product quality, customer satisfaction, corporate image, factory safety, employee satisfaction, and environmental impacts.

d. Economic Life. It is also difficult to predict how long an item of equipment will continue to function. Even if equipment life could be determined accurately, there would be the problem of annual cost and benefits changing (generally getting worse) as the equipment ages and wears out.

The term *economic life* is preferred to *useful life*, because an item of capital equipment is rarely used to the point at which it is no longer useful. Generally, it is used only so long as keeping it is the most economical alternative. As a piece of equipment ages, its maintenance costs almost always increase. Further, the depreciation allowance eventually gets used up. At some point, it is advantageous to purchase a newer piece of machinery, even though the old one still has some life left in it.

The point at which this replacement becomes the preferred alternative is very difficult to determine, especially at the beginning of the life cycle of the original equipment. The future maintenance costs are difficult to anticipate, as are the future costs of replacement equipment.

The point of this discussion is that any estimate of the economic life of a capital investment must be considered to be a very rough approximation. The only exception is when you have already been through many life cycles of the same type of item and have extensive experience to draw upon.

e. Salvage Value. This is another value that is difficult to determine at the beginning of a machine's life cycle. It will vary with the economic life of the equipment, the final condition of the equipment, and the prevailing economic, technological, and market conditions at the time of replacement.

f. After-Tax Cash Flow. Once all the values discussed above have been estimated, as closely as they can be, an economic analysis is possible. However, this will be a before-tax analysis, which does not necessarily reflect the actual cash flow that will result.

In general, money spent in the course of doing business is tax deductible, but there are several different ways of deducting it. If the items purchased will be consumed within a year, they are considered ordinary expenses. These expenses include office supplies, rent, salaries and wages, and insurance. The costs of these items are deductible in the year that they were purchased.

Other items last more than one year. These are considered capital expenses, and their costs cannot be deducted in a single year. They include such items as machinery, computers, buildings, vehicles, and land. The costs of these capital items must be deducted over the course of several years, and each year's deduction is called a *depreciation allowance*. A large number of laws and regulations spell out detailed rules for determining how many years it takes to fully depreciate each type of equipment and how much depreciation is allowed in each year. These laws are complex and tend to change from year to year.

In conducting a complete economic analysis of any investment under consideration, it is necessary to calculate the tax consequences, including depreciation allowances. The net effect of taxes is to minimize both profits and losses, since payment of the taxes will remove some of the income, and depreciation allowances will return some of the expenses. These after-tax rates of return are the numbers that must be compared among the various alternatives to actually determine the most economically attractive course of action for an investment.

IV. SUMMARY

This chapter addressed manufacturing planning, beginning with the strategic manufacturing management element of TMA. Strategy planning plays a vital role in defining where a corporation wants to be in the near and long terms and how it will get there. The connection with TMA is clear; TMA must be an integral part of the overall corporate business strategy.

A strategic manufacturing plan consists of two fundamental pieces of information: strategy and tactics. Strategic planning involves assessing the internal and external worlds that affect the corporation. This is accomplished via analyses addressing areas such as the environment, competition, investment, potential, distribution, and customer base.

A key consideration in strategy development is the product itself. The product must be able to fill a market void and differentiate itself from the competition. Key performance parameters are reliability, safety, and quality. It makes sense to develop a strategic manufacturing plan that optimizes these parameters to increase the probability of successful commercialization.

Tactical planning involves defining the methods for guiding, supporting, and performing the actual manufacturing activity. Topical methods such as design for manufacture, computer-integrated manufacturing, material requirements planning, just in time, and economic analysis are very important and beneficial.

V. QUESTIONS

1. Identify a broad range of topics that could be addressed in a business strategic plan.
2. Business strategic plans typically begin with a mission statement. Develop a concise, explicit mission statement for your area of concern.
3. Identify the external and internal issues impacting your corporation or current activity.
4. Discuss the significance that environmental issues might have for some corporations in formulating strategy.
5. Discuss the differences between tactical and strategic planning.
6. Select some simple item that you can disassemble (pencil sharpener, stapler, etc.). Suggest ways to redesign the item in keeping with the guidelines for DFM.
7. List three advantages of CAD that your company would benefit from.
8. Discuss the difference between an FMS and an FMC.
9. Describe two benefits to be gained by increasing employee understanding of many parts of a company's production processes.
10. Describe the relationship between time buckets and planning horizons.
11. Where do the inputs to the basic MRP equation come from?
12. Is a lot size of less than one at all similar to the simultaneous production of mixed product lines? Why or why not?
13. Discuss several ways to implement a pull system for a production line in your factory.
14. What is the difference between inflation and the "time value of money"? Are they related at all?

VI. REFERENCES

1. Buzzel, R. and Gale, B.: Market Share—A Key to Profitability. *Harvard Business Review*, 1975.
2. Hunger, J. and Wheeler, T.: *Strategic Management And Business Policy.* Addison-Wesley Publishing Company, England, 1986.
3. Aaker, D.: *Developing Business Strategies.* John Wiley and Sons, New York, 1988.
4. Tersine, R.: *Production/Operations Management—Concepts, Structure, and Analysis.* North Holland, New York, 1980.
5. Boothroyd, G., Poli, C., and Murch, L.: *Automatic Assembly.* Marcel Dekker, New York, 1982.
6. Groover, M.P.: *Automation, Production Systems, and Computer-Integrated Manufacturing.* Prentice-Hall, Englewood Cliffs, New Jersey, 1987.
7. Vollman, T., Berry, W., and Whybark, D.: *Manufacturing Planning and Control Systems.* Richard D. Irwin, Inc., Homewood, Illinois, 1988.

8. Hernandez, A.: *Just-in-Time Manufacturing*: *A Practical Approach*. Prentice-Hall, New York, 1989.

9. Newnan, D.G.: *Engineering Economic Analysis*, third edition. Engineering Press Inc., San Jose, Calif., 1988.

10. DeGarmo, E.P., Sullivan, W.G., and Bontadelli, J.A.: *Engineering Economy*, 8th ed., Macmillan Publishing Company, New York, 1989.

4

Management Control

As in any corporation, management control must be maintained and enhanced wherever possible. This does not mean overmanagement, but clear, effective, and efficient management that motivates and empowers people.

This chapter first addresses the need to establish an integrated approach among various departments and people to achieve TMA, as well as ways to do this. Then, recognizing that people and departments have their own personalities, we discuss the need to establish and implement a motivational strategy that promotes a positive working environment for *all* persons.

The next section addresses various qualitative and quantitative engineering management tools. The last section illustrates the power and flexibility provided by expert knowledge-based systems. Expert systems are an excellent way to increase management efficiency and eliminate or reduce management and operator error.

I. INTRODUCTION

Management control and leadership are critical factors in the success of a corporation. Accepting this, we recognize that both control and leadership are an integral part of the whole TMA objective. TMA is not intended to be a trendy or fashionable activity that makes a grand debut and then quickly disappears, but rather a long-term commitment to continual improvement as a leading corporation.

Control is demonstrated by maintaining a direct and expedient course toward corporate strategic goals and objectives. Leadership is demonstrated through the desire to ensure that products are well designed initially and then manufactured and provided to the customer without any loss of design integrity.

This chapter outlines several concepts and tools to facilitate management control of TMA activities. These concepts and tools are not presented as new and innovative techniques, but as underused and very effective and efficient ways to maintain management control.

II. INTEGRATED MANAGEMENT CONTROL

Chapter 3 introduced many of the important elements of a manufacturing environment that are used to attain TMA. To be effective, these individual elements must work together, communicate, feed on each other's strengths, and shore up each other's weaknesses. This requires that they be implemented through an integrated management control approach.

For this integrated structure to exist, management must take a very strong hand in all areas of endeavor, including engineering, manufacturing, marketing, and product assurance. Management must have a wide-ranging vision and avoid mistakes of shortsightedness or narrowness of focus. Far more than in traditional manufacturing environments, the TMA-oriented corporation must be glued together by the integrating powers of its management function.

This is not to say that successful TMA requires more management than traditional organizations. In fact, the truly integrated company often has fewer layers of management and fewer people employed at each level. This is not the contradiction that it seems to be at first glance. Integration of function requires that all hands know what each of the other hands are doing, and this becomes more natural when there are fewer hands involved. A large management structure tends to specialize and lose broad focus; a lean management structure necessitates a broad overview on the part of all participants.

The motto of the TMA management structure is "no walls." All departments and functions must cooperate and share information, responsibility, and decision making. Interdisciplinary teams must be a standard working paradigm at all levels of control.

Allowing walls between various functional groups encourages a dangerous "throw it over the wall" style of interactions. In this mode, marketing will decide what product is needed, write it in a memo, and "throw it over the wall" into the design engineering department. Design will come up with what it deems to be an acceptable product design and then will toss the blueprints over the wall to process engineering, which will come up with its vision of a good process plan. This will be lobbed over another wall to manufacturing, and so on.

What the organizational walls actually do is to force a sequential or serial operation and communication structure (see Figure 1a). This limits interactions, inhibits learning from the experiences of other groups, and generally keeps everyone's focus entirely too local for TMA.

More appropriate is a parallel operation and communication structure, in which multiple departmental tasks are carried out more or less at the same time (see Figure 1b). Because certain functions must naturally lead into other functions with their output, complete parallel structure is not usually feasible. But the more cooperation and interaction that is possible, the stronger the entire organization.

As mentioned above, an integrated management approach requires a broad focus on the manufacturing enterprise. Certain management guidelines make this philosophical strategy more concrete:

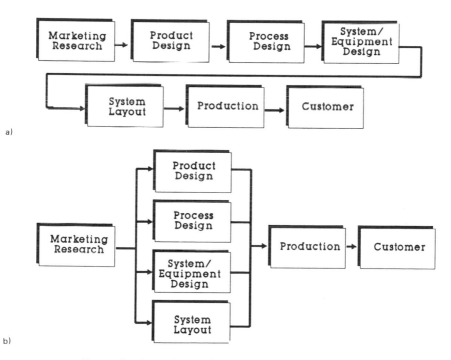

Figure 1 Operation and Communication Structure

1. Focus on broad, rather than narrow, performance measures. Avoid judging manufacturing tactics on short-term and short-range indices of performance. These shortsighted measures include such ideas as machine utilization, labor hours per part, floor space, inventory levels, and so forth. This does not mean that these goals are bad or even unworthy; they are quite good. But they are not proper goals for management to focus upon.

More fitting goals for high-level management are competitiveness in the industry, corporate image, customer satisfaction, adaptability to changing market conditions, and other issues that provide long-term staying power in relation to competitors. These high-level goals must be the guiding principles. When they are taken care of, the narrow-focus measures will take care of themselves.

2. View change as a process, not as an event or program. In keeping with a broad focus, recognize that change is the only constant in the manufacturing business, as in life. It is a mistake to look at a new production line or new organizational structure as a sudden change, dividing the old from the new in a quick, clear-cut delineation. Rather, see the gradual implementation of new techniques and equipment as a way of life that, when constantly followed, ensures constant improvement.

3. View reliability, safety, and quality (RSQ) as a solution, not as a problem. Too often, management sees acceptable product functional performance as the goal and sees RSQ as an impediment to that goal. This philosophy leads to a manufacturing environment that implements only enough RSQ to avoid missing the given specifications. It also delays the concern with the RSQ of products and processes until after they have been both designed and implemented. It results in a case of too little, too late, and undermines the competitive edge.

The correct philosophy is to see RSQ as a primary goal, and satisfaction with meeting specifications as the problem. When all efforts are tuned toward maximizing RSQ in all respects (including design, process, and testing), the best possible, most competitive output will result.

4. Use shorter time frames than seem comfortable. Typical time frames for corporate programs are in the range of three to five years. The goals of these programs may be the implementation of a new customer service department, the complete automation of an old process, or perhaps improvement of some performance index by a given percentage. Do not be afraid to attempt the same goals in half the usual time, or even less. John Kennedy set the seemingly impossible goal of putting a man on the moon before the end of the 1960s, and the goal was met. Forcing an organization to react as quickly as possible can motivate it to react as efficiently as possible.

5. Develop a preemptive attitude toward problems. Do not merely wait for problems to occur, thinking that they can be fixed when they crop up.

This attitude is tempting, since it requires action only on problems that do, indeed, develop. But it is self-destructive and anti-TMA in the long run. The broad-focus approach is to develop an organizational structure that automatically senses any process that is beginning to deviate even slightly and forces a solution before it can develop into a full-fledged problem. Just-in-time is an example of a preemptive approach to production control (see Chapter 3).

6. Be willing to take risks. There is an entire world full of competitors out there, and many of them are going to be taking risks. Some of those that do will fail, and fall behind, but some of them will succeed, and leave you to fall behind. To remain competitive, you must take risks, and you must make them pay off.

Refusing to take risks is like the poker player who always antes, but never sees a bet. He will get nickeled and dimed to death and will never take home a pot.

These philosophical concepts may help develop broadly focused viewpoints for top management to follow, but they do little to help organize a corporation that is capable of following them. The successful organization is one with the proper interlinking of functions and lack of walls to allow these concepts to be implemented.

Guidelines exist for proper structure, some of which have already been discussed:

1. Structure like a matrix, not like a pyramid. The typical organizational chart is based on a two-dimensional pyramid, with large numbers of low-level functions, each feeding upward to smaller and smaller levels of control. The top level is completely isolated from the lower levels, and no sideways communication is possible.

To eliminate walls, this structure should be rearranged into a multidimensional matrix. The pyramid should be flattened so that there are fewer layers from the top to the bottom. The size of the bottom layer should be decreased and that of the top layer increased, so that workers are less specialized and top managers are less abstract. There should also be a large number of horizontal lines of communication, so that all functions interact appropriately in all phases of a project. This fits in with the parallel, rather than serial, structure discussed earlier.

2. Develop interdisciplinary knowledge in all personnel. Just as manufacturing functions are integrated by computer linkages and nontraditional organizations, people must also be integrated. This should be accomplished at two levels: the level of the individual and the level of the group or department.

Encourage, or even require, individual employees to learn as much as possible about other departments and their functions, responsibilities, and problems. Publicize open positions within the company, and encourage employees to apply for them. Institute training periods for new employees or newly promoted managers that rotate them through a variety of functions.

At the group level, organize multidisciplinary teams to attack problems. Draw members for key committees from a large variety of departments. Invite people to meetings from different functional areas of the company. Whenever a group is organized, try to get a cross section of the corporation's expertise involved.

3. Eliminate barriers to communication. Often, different departments have trouble communicating because of differences in vocabulary, viewpoint, or educational background. Try to standardize how information is stored and presented throughout the organization.

Vocabulary can be a significant problem. Each department may have a different term for the same concept or may use the same term for different concepts. Try to root out these sources of confusion and develop a standard set of definitions.

Time frames are another area where separate functions have trouble communicating. Top managers tend to think in terms of years, while sales and marketing personnel tend toward months and quarters. First-line managers think in terms of weeks and months, while workers often focus directly on individual shifts, or even hours. Although it is probably not possible to get everyone to agree upon a single unit of time, at least translate each individual piece of information into the appropriate time frame for the target audience.

4. Avoid the "not-invented-here" (NIH) syndrome. This is the attitude that if some other department uses this approach, we don't want it, we'll come up with our own. This attitude wastes incredible amounts of time, as each functional unit attempts to reinvent its own wheel. Try to cultivate a team attitude; all departments and divisions are on the same side.

Management not only must prevent the NIH syndrome from wasting resources within the company, it also must resist the urge itself. Evaluating and possibly enhancing ideas derived from other companies may seem repugnant, but it is essential to retaining competitiveness.

5. Do not allow stagnated benchmarks. Posting a long-term stationary goal for a specific function can only lead to complacency and stagnation. Keep goals moving and constantly updated. Better yet, set metagoals, such as increasing the amount of improvement each month, or going longer without a defective part than the last time.

III. MOTIVATION

A fundamental problem in organizations is complacency. This results when people are no longer motivated to advance the well-being of the corporation or themselves. Complacency makes it very hard to maintain a progressive and aggressive corporate attitude.

Complacency should also be regarded as an infectious disease. If too many people catch it, it can cause a corporation to go to sleep, which can have catastrophic effects. The key to combating complacency is to motivate and empower people. If this is not possible, complacent people must be removed from the organization. The bottom line is that top corporate management is responsible for providing the leadership necessary to motivate workers to function in a state of high creativity, entrepreneuralism, and productivity.

In Chapter 1 we stated that if a person is unhappy with his job, he should look for a new job. A complacent attitude may be the result of a person being too happy with his job. That is, he is happy with the status quo. An objective of effective leadership must be to motivate people not just to be happy with their jobs, but to continually look for ways to improve them and make them more rewarding both personally and for the corporation.

Ideally, every person should intrinsically enjoy making an effective contribution toward corporate growth and profitability. However, although all workers are capable of self-motivation, management generally must provide the motivational spark through a proper leadership style. The leadership style by which motivation is achieved must remain flexible to accommodate change, as necessary, so that the organization as a whole can avoid the complacency trap. This is particularly true in light of the broad range of personalities, educational backgrounds, and organizational levels contained within a corporation.

A. Leadership Style

Four factors are identified as being associated with effective leadership [1]. First is *support*, which involves making people feel worthwhile and aware of management consideration. Second is *interaction facilitation*, which involves fostering the development and effective use of people and establishing a good communication network. Third is *goal emphasis*, which involves making people aware of the goals of the organization and instilling a sense of enthusiasm for achieving these goals. Fourth is *work facilitation*, which involves helping people achieve the corporate goals by effectively administering, planning, and coordinating resources.

These four factors are certainly important and should be employed fundamentally at every level of organizational leadership. However, they must be applied with a perspective on the specific workers being targeted. For example, it likely that exactly the same leadership style will not be effective for workers on the production floor and workers in the corporate office. These two sets of people have different personalities and divergent corporate expectations.

It is clear that to meet corporate goals, including those that are TMA specific, it is important to determine the appropriate leadership style required. This must be done at the corporate level, as well as at each successive organizational management level.

In general, three variables are considered in determining the appropriate leadership style: (1) forces in the projected leader, (2) forces in the other organizational members, and (3) forces in the situation [2]. The best leadership style depends on management values, personality characteristics, and feelings in various situations. Further, it depends on whether people are ready to accept responsibility, are interested in company problems, and are in tune with the corporation's goals.

These three variables can be considered together using Vroom's prescriptive model of appropriate leadership style [3]. The leadership style serves as the cornerstone of the motivational strategy. The model is exercised using a decision logic that defines five alternative leadership styles, as identified below and keyed to Figure 2.

1. Decisions are made solely by the manager based on whatever information he or she possesses.

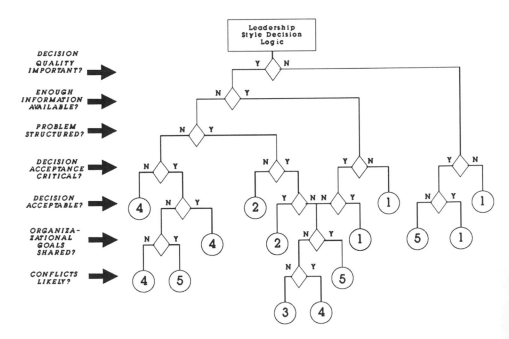

Figure 2   Leadership Style Decision Logic [Gilmer and Deci/Industrial and Organizational, Psychology, 4E (1977) McGraw-Hill, by Permission.]

2. Decisions are made solely by the manager, but additional information is gathered from workers.
3. Decisions are made by the manager, but significant input is derived from workers on an individual basis.
4. Decisions are made by the manager, but significant input is derived from workers on a group basis.
5. Decisions are made by the manager *and* the workers as a group.

By exercising the decision logic, the optimal management leadership style is defined.

B. Organizational Psychology

Organizational psychology is a broad and very interesting field. Much research has been performed and the field continues to be an attractive research area. Several prominent organizational behavior models and motivational theories have been defined. These provide a foundation for establishing corporate, divisional, and departmental motivational strategies.

There are four important organizational behavior models [4]: (1) autocratic, (2) custodial, (3) supportive, and (4) collegial. The autocratic model is founded on the power base of the manager and strict compliance of the worker. It was most prevalent over eighty years ago. In the 1920s and 1930s it yielded to the custodial model, which proved more successful. This model is founded on the economic resources made available to the worker and on his or her satisfaction. The supportive model is widely accepted and predominates in many organizations. This model relies on leadership and the subsequent motivation of the worker. The collegial model requires mutual contributions from both management and the work force. The result is a firm commitment to the job at hand and to the organization. Currently, many progressive organizations use the collegial model.

Some of the most widely known motivational theories include those presented by Maslow, Vroom, McGregor, and Herzberg. Maslow's hierarchy theory states that people progress through a hierarchy of needs. These consist of physiological needs, safety needs, social needs, esteem needs, and self-actualization needs. Each level of needs provides a source of motivation. Once a level is satisfied, the person is motivated to achieve the next level.

Vroom defines the Expectancy, or Path-Goal, Theory. He states that motivation depends on what a person expects to receive for his or her efforts. This expectation is not limited to first-order needs such successful task completion, but includes second-order needs to which the first-order needs might lead (for example, a pay raise).

McGregor defines the famous Theory X and Theory Y notions. Theory X states that workers are lazy and will work only to the extent that their work is instrumental to getting rewards. Theory Y, on the other hand, states that workers are not lazy but are self-motivated to set and achieve their own goals, which are generally consistent with the goals of the corporation.

Herzberg defines the Motivation-Hygiene Theory. This theory states that people have two sets of needs. The first set includes needs related to the job context: salary, working conditions, supervisory policies, and so forth. People are never satisfied by these factors, but are merely neutral. Their absence leaves people dissatisfied. The second set includes needs related to the content of the job: challenge, the need to use one's full capacities, the need to be resourceful, and so on. The absence of these needs leaves people neutral, and their presence makes people satisfied.

There are many more motivational and organizational behavior theories. Those identified here are only a sample of some of the more widely known. The reader is encouraged to explore this area of psychological research. One point should be obvious: no one theory provides all the solutions for all possible situations. Corporate cultures are complex phenomena and require a motivational strategy based on a hybrid of the numerous theoretical approaches defined.

C. Strategic Motivation Model

The nature of most organizational activities allows them to be depicted by a work path to achieve an end goal; this is characteristic of Vroom's Expec-

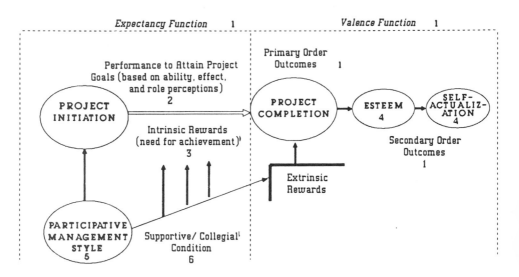

Figure 3   Strategic Motivation Model [5, 7-11]

tancy Theory [4]. From this baseline framework, other theoretical ideas fit easily around Vroom's notion (see Figure 3). Together, these ideas form a strategic motivation model.

As shown in the figure, the basic project path and ultimate goal are functions of what is expected to be accomplished and the value of the accomplishment. Once an organizational activity is initiated, individual responses are based on ability, effort, and role perceptions. An intrinsic need to complete the activity successfully exists within the organizational team members. To foster this need throughout the activity's duration, a supportive and collegial condition is defined that provides the greatest opportunity for creativity in attaining the end goal. Of course, the primary goal is successful completion. Derived from attainment of this goal is monetary compensation, as well as self-esteem and, ultimately, self-actualization.

This strategic motivation model is intended to provide the most suitable environment for the organizational members it affects. It is designed to emphasize both individual and group creativity to further a modern, innovative, and progressive corporate attitude.

IV. QUALITATIVE AND QUANTITATIVE TOOLS

Throughout the daily operations of a manufacturing organization, there are a multitude of management decisions being made in conjunction with many ongoing activities. This is true for corporations that are working for TMA, as well as for corporations that are not striving for excellence. It is imperative that informed decisions be made and that decisions be in accordance with a defined strategic theme. To ensure that this is the case, it is essential that formal project management exist as part of the corporate culture.

The key to successful project management is up-front planning. For a given project, the activity sequence, requirements, and major milestones (for example, proof-of-concept and technology transfer) are defined and integrated into a basic project management plan. Various qualitative and quantitative analysis techniques are then applied to adjust the plan as necessary. The result is a plan that reflects the optimal approach to achieving the end goal(s) or objective(s) in support of TMA.

The fundamental objective of management discipline is to enhance the benefit-cost-time potential of a project. Without a defined project management activity, there is a danger that little or no coordination will occur among all the participating groups. There is nothing worse than thinking a project is completed, only to find that fundamental elements of the project slipped through the cracks or that the completed elements fail to fit together.

For example, consider a product development project. It is known that successful commercialization depends on the product meeting or exceeding its design performance goals, including reliability, safety, and quality. Suppose that the product, despite having undergone a successful development program, lacked commercial feasibility due to a lack of some fundamental performance characteristics required by the market. Somebody failed to get marketing involved to determine exactly what the targeted customer wanted. The preparation of a project management plan that identified all tasks and communication interfaces would have eliminated, or greatly reduced, the risk of such an error.

Sound project management optimizes the overall effectiveness of a project in regard to its constituent activities. With this approach, one minimizes the risks associated with a large number of activities and participants, and maximizes their contribution to effectiveness. Project management also ensures that the proper activities are planned, that the activities are properly timed, that adequate resources (e.g., personnel, test items, test facilities, funding) are available, and that no unnecessary or redundant activities are conducted.

As a minimum, a project activity schedule should be developed. This involves defining the major project tasks and subtasks that will be performed. Also, a timeline and milestones are defined for each task. Figure 4 illustrates a candidate form for documenting this information.

Typically, the effectiveness of a project management plan is optimized with respect to estimated cost and time. This enables rational choices to be made with respect to overall project content and structure based on a return-on-investment approach.

There are many techniques available to aid optimal management. Some of these include:

- Networking
- Linear programming
- Inventory, production, and scheduling
- Econometrics, forecasting, and simulation
- Integer programming
- Dynamic programming
- Stochastic programming
- Nonlinear programming
- Game theory
- Decision theory
- Optimal control
- Queuing
- Difference equations

All these techniques have their place in practice. Their common objective is to help in making the best, or optimal, decision. The top two items in the list are frequently used by industry and are discussed in further detail in the following subsections.

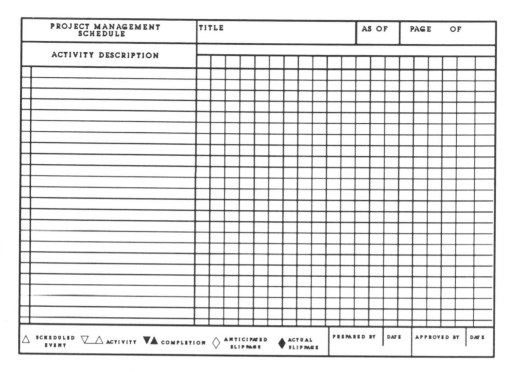

Figure 4   Project Management Schedule Form

A. Project Management Networking

A good tool for defining and guiding the project management plan is the use of a networking technique [12]. Networking is an enhanced graphical application of the program evaluation review technique (PERT) and the critical path method. The network provides a view of the interactions and interconnections of various project activity time constituents. It also provides a powerful tool for identifying and recognizing relationships among interdependent activities.

This is particularly important because completing projects on time and within budget constraints is often difficult at best. The fact that certain activities must be completed before others can be started complicates matters. Therefore, the ability to deal effectively with the interdependency among activities provides a great advantage in achieving success.

A network itself consists of a series of ovals, or nodes, connected by lines. The nodes represent the completion of activities and attainment of a project milestone. The nodes themselves contain several pieces of information. These include (1) the activity reference number, (2) the earliest start time, (3) the latest start time, (4) the earliest finish time, (5) the latest start time, and (6) the activity duration time. Figure 5 illustrates the components of a network node.

The lines between each node represent the activity in progress. The length of each line has no correlation to the actual duration time of the activity and is merely a convenience for the user.

The network illustrates how all the tasks of the project are tied together. The saying "a picture is worth a thousand words" certainly applies to project management networks. The network identifies which activities must be performed before others, which occur in parallel, and which are unrelated. In addition, the time required to complete individual activities, as well as the overall project, is clearly identified.

An important early step in preparing a network is to define the activities required in the project and to establish the proper order of precedence. Errors or omissions at this step will produce a faulty management plan and possibly have a catastrophic effect on the project at some point.

Proper project management networking involves identifying each activity, its description, its immediate predecessors, its estimated duration, and the resources available. The immediate predecessors of an activity are those activities that must be completed before the start of the activity in question. The estimated duration time is defined by discussion with the various persons participating in the project. Typically, both the overall time and the resources available are well-defined project constraints and reflect corporate priorities.

As an example of project management networking, consider a project to conduct a design verification test for a piece of hardware that is part of an overall system. The following test activities are defined as essential parts of the corresponding network diagram presented in Figure 6:

1. Test unit fabrication
2. Special test equipment development and fabrication
3. Test procedure preparation

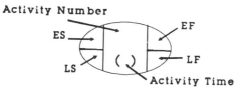

Figure 5   Components of a Network Mode

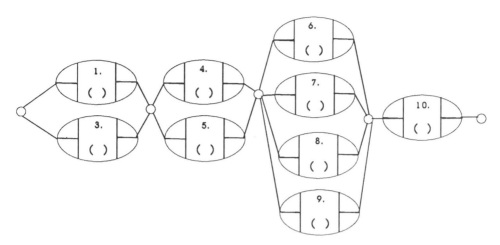

Figure 6   Example Project Management Network

4. Support requirements definition and establishment
5. Test configuration setup
6. Test conduct and monitoring
7. Test data compilation
8. Failure reporting, analysis, and corrective action
9. Test facility use
10. Test data analysis and reporting

The network depicts the activities in their order of precedence. Note in the diagram that some of the project activities occur simultaneously, as would be expected.

Once the project manager develops a network diagram, it is then filled out by estimating the time required for each activity. For all network diagrams, the earliest and latest start and finish times are defined to determine the critical path [13] and, subsequently, the total calendar project time required.

The earliest start time (EST) is the earliest time an activity can begin, when all preceding activities are completed as rapidly as possible. The earliest finish time (EFT) for each activity is the EST plus the activity time. When two or more parallel activities are immediate predecessors to another activity, the EST for the activity is the largest of the EFTs of its predecessors. The EST and EFT are defined by starting with the first activity and working through the network to the final test activity.

The latest finish time (LFT) is the scheduled time when the test program must be completed. This time is fixed independently of the network diagram and, typically, corresponds to market introduction factors. The latest start time (LST) is the latest time at which an activity can be started if the test program schedule is to be maintained. The LST is LFT minus the activity time. These times are defined by starting with the final activity and working backward through the network. To determine the LFT for an activity that has two or more parallel successors, the smallest LST of its successors is selected.

The EST and EFT times are developed by a forward pass through the networks. Identification of the LSTs and LFTs is obtained by a backward pass through the networks. The total slack is the difference between the two start times *or* the two finish times (these times are the same for each test activity). All activities along the critical path have the same amount of slack. Slack will either be zero or some minimum value for all the activities. The critical path activities are crucial to the on-time completion of the test program. These activities receive the greatest amount of management attention.

In some cases, it may be found that the critical path is too long and must be shortened. To do this, two basic approaches are used. The first is the strategic approach. This consists of questioning the defined order of precedence of activities, particularly those on the critical path. It may be possible to make arrangements to complete some activities in a manner that removes them from the critical path (i.e., in parallel with some other activity).

The second is the tactical approach. This consists of determining if an increase in or acceleration of funding resources reduces the time of certain activities in the critical path. Initiating the paid working of overtime is an example of this. This approach requires a trade-off between project time and cost.

B. Optimization Modeling

As alluded to in the previous subsection, decisions are typically made in light of some definable constraint(s). Constraints many be self-imposed or dictated by others. Most frequently, constraints are a function of time and money.

To make the best decision, it is necessary to balance, or trade off, project variables (e.g., time and money). Constrained optimization modeling (COM) is used to make management decisions that achieve the best possible results, within given restrictions. The whole area of making optimal decisions is encompassed in the field of management science. One of the most popular management-science tools used for COMs is linear programming.

Linear-programming-based COMs are used in defense, health, transportation, energy planning, and resource allocation. In purely private-sector applications, uses include long-term and short-term scheduling of activities. Long-term planning activities include topics such as capital budgeting, plant location, marketing strategies, and investment strategies. Short-term planning activities include production and work-force scheduling, inventory management, and machine scheduling.

In all applications, the objective is to make optimal decisions that maximize some desired outcome. The trick for each application is to satisfy the defined set of restrictions or constraints in arriving at an optimal solution.

Note the continual reference to the "optimal solution." A COM provides a solution that is called the optimal, or best possible, answer. Remember that these optimal solutions are produced relative to a mathematical problem posed by a model. Therefore, the optimal solution is only as good as the model defined to described the problem.

Keep in mind that "optimality" is a theoretical concept, as opposed to a real-world concept. Although it is advantageous to have quantitative data to support and guide decisions, this alone should not form the basis for decisions. Management and engineering intuition play vital roles in making sound, well-rounded decisions. Likewise, COMs are of great help in making final decisions, particularly in identifying the numerous variables involved in making a specific decision. However, the solutions generated must not be used in a vacuum that excludes intuition.

1. Linear Programming

In structuring a COM, one must identify all the mathematical variables required in making the decision. The more complex the model, the more difficult it is solve. However, increasing complexity also makes the model's solutions more accurate and credible. The bottom line is that models which realistically describe a situation are most apt to provide the best answers.

Let's look at the structure of a COM solved by linear programming. Every COM consists of an objective function, decision variables, constraints, and parameters.

The objective function describes a decision function. To derive an optimal solution, the objective function must be either maximized or minimized relative to a set of constraints. Making up the objective function are the decision variables. The decision variables represent actions or activities to be undertaken at various levels. Any selection of numerical values for the decision variables indirectly assigns values to the constraint functions. The COM requires that each of these constraint function values satisfy a condition expressed by a mathematical inequality or equality. The information serving as limiting values are the parameters.

There are five key steps in defining a COM [14]:

1. Express each constraint in writing, noting whether the constraint is a requirement (of the form $>$), a limitation (of the form $<$), or exact (of the form $=$).
2. Express the objective (i.e., maximize or minimize something) in writing.

3. Identify the decision variables.
4. Express each constraint in terms of decision variables.
5. Express the objective function in terms of decision variables.

In structuring the COM it is important not to read more into a problem than precisely what is known.

Linear-programming-based COMs can rapidly become very complex. Such problems are conveniently solved using a computer to arrive at the optimal solution. (Numerous commercial software programs are available.) It is possible, however, to solve simple linear programming models manually.

The simplest COMs are composed of two decision variables. A simple COM allows a graphical solution and enhances the conceptual visualization of a particular management problem. Graphical solutions provide key advantages in evaluating different problems. One sees what happens visually by adjusting the constraints. This might include adding a constraint to a problem, tightening or relaxing a constraint, unintentionally leaving out a constraint, or including a constraint in the model in the wrong way.

Linear programming uses the fact that equations of the form $ax + by \geqslant c$ are linear inequalities. This means that all constraints or restrictions in the COM are linear inequalities. When the above equation is changed to $ax + by = c$, it becomes a linear equation. In general, equations in which one side is of the form $ax + by$ and the other side is a variable value that depends on the choice of x and y are called *linear functions.*

Knowing that the graph of any linear equation is a straight line, we can draw this line merely by locating two points in the plane that lie on the graph. This is accomplished by finding the points of the line that intersects the x and y coordinate axes. To do this, a zero value is substituted for x to find the value of the y coordinate, and vice versa. This gives a set of two points—that is, $(0, y)$ and $(x, 0)$—which graphs the linear equation.

This is done similarly for each constraint. The graph of all constraints together defines a domain, or feasible region, which isolates all the possible combination of variables satisfying the objective function (see Figure 7).

The domain for the COM is determined by the relevant side of each constraint's equality line. This relevant side is determined by selecting a trial point (e.g., 0, 0) that is not on the constraint line. If the trial point satisfies the original inequality, then the linear line plotted and all points on the same side of the line as the trial point satisfy the inequality. If the trial point does not satisfy the original inequality, then the linear line and all points not on the same side as the trial point satisfy the inequality.

Depending on whether one desires to maximize or minimize the objective function, the optimal variables are identified as the extreme points of the domain. These points are represented by a corner point of the domain; that is, a point where two edges of the domain come together. Note that the objective

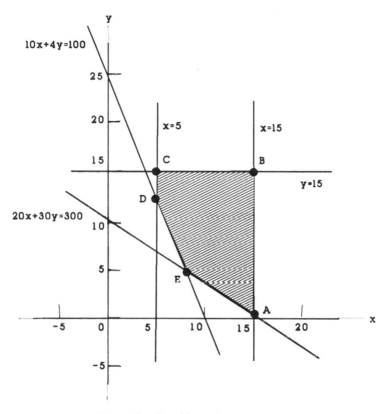

Figure 7   Graphing of Constraints

function is also a linear function and its graph passes through the extreme points of the domain for the optimal solution.

To sum up, linear programming optimization problems involving two unknowns are solved by three fundamental steps:

1. Construct a geometric representation of the domain, or a feasible solution set of the optimization model.
2. Identify the extreme points of the domain.
3. Calculate the value of the objective function satisfying the criterion for maximization or minimization.

Keep in mind that optimization problems involving only two variables are unique. It becomes difficult, if not impossible, to solve a COM involving several variables through geometric methods. A popular alternative to geometric solution methods is the Simplex Method.

The Simplex Method is a technique of matrix manipulation that automatically finds pairs of linear equations with graphs that intersect at extreme points, solves for variable values that satisfy all constraint equations, and calculates the values of linear functions for the variable values found. In addition, the Simplex Method identifies the variable value combinations that solve the linear programming problem and, in general, does not need to examine all the extreme points to do this.

The most important feature of this method is that it works for linear programming problems in any number of unknowns, not just two. Also, it is well suited to computer implementation in solving large linear programming problems. As with all decision techniques, however, there are some applications not suited for the Simplex Method. Nonetheless, most COM-based linear programming problems arising in practice can be solved using this method.

2. COM Example

Consider the need to establish the most effective product development test program plan. To do this it is necessary to find the optimal solution through geometric representation of a linear-programming-based COM. A graphical method is used because the natural segmentation of a product development test structure (i.e., laboratory testing and field testing) allows for two-dimensional analysis. It is desired to determine the maximum effectiveness levels for laboratory testing and field testing based on various project constraints.

We define the following objective function:

maximize $70EL + 30EF$

where EL is the effectiveness level for laboratory testing and EF is the effectiveness level for field testing. The objective function is derived based on a maximum overall program effectiveness limit of 100 percent.

We also define the following constraints:

1. $EL, EF \geq 0$
2. $EL, EF \leq 1.0$
3. $70\ EL + 30\ EF \geq E$
4. $tL\ EL + tL\ EF \leq T$
5. $cL\ EL + cF\ EF \leq C$

where the decision variables are defined as follows: E is the minimum acceptable effectiveness level defined for the overall project (see Table 1) and the 70 and 30 represent effectiveness contributory weights; tL is the estimated laboratory test time required, tF is the estimated field test time required, and T is the actual test program time available; and cL is the estimated laboratory testing cost, cF is the estimated field testing cost, and C is the actual R/D test program funding available.

Table 1 Test Plan Effectiveness Levels

Level A—Highly Effective:
  A very high probability is required of the test program plan to contribute to successful technology transfer. This corresponds to a range of 100 to 85.

Level B—Effective:
  A significant probability is required of the test program plan to contribute to successful technology transfer. This corresponds to a range of 84 to 70.

Level C—Reasonably Effective:
  A moderate probability is required of the test program plan to contribute to successful technology transfer. This corresponds to a range of 69 to 55.

Level D—Remotely Effective:
  A low probability is required of the test program plan to contribute to successful technology transfer. This corresponds to a range of 54 to 30.

The COM defined states that the problem is to make the value of the objective function as large as possible, provided that the constraints are satisfied. The value of the objective function is measured in effectiveness (i.e., 0 to 100). Constraint 1 defines the nonnegativity conditions. Constraint 2 defines the upper bound of laboratory and field testing effectiveness. Constraint 3 defines the test segment effectiveness level possible based on the lower acceptable bound of overall test program effectiveness and the base line test segment effectiveness weights. Constraint 4 defines the test segment effectiveness level possible based on the maximum test program time available and the program time identified for each test segment. Finally, constraint 5 defines the test segment effectiveness level possible based on the test program funding available and the estimated cost of each test segment. The model requires that each of these constraint values satisfies the condition expressed by the mathematical inequalities defined.

The established constraints are solved and plotted to define a feasible region from which the optimal corner is found and, consequently, to define the optimal effectiveness levels for laboratory and field testing. Figure 8 illustrates a possible feasible region formed by a constraint set. The optimal corner determined by the constraint set is the one that maximizes the objective function defined previously.

V. EXPERT SYSTEMS

One of the newest and most powerful tools available for management control of manufacturing is the expert system. In fact, this tool is important enough to devote a separate section to it, although it could easily have been

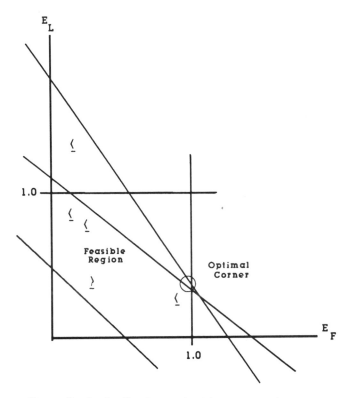

Figure 8   Optimality Determined by a Constraint Set

incorporated into the previous section. The power and flexibility of the expert system concept, however, sets it apart from the algorithmic tools discussed previously.

A. Expert System Definition

An obvious question is "What is an expert system?" Many definitions exist, and not all are completely compatible. For TMA purposes, an expert system is defined as a computer program that relies on (1) knowledge and (2) reasoning to heuristically arrive at a decision or perform a task that is usually accomplished by a human expert.

It is important to recognize that this definition sets expert systems apart from the many other types of computer programs. Several concepts are implicit in our definition, each of which leads to a specific strength of the expert system approach.

- Knowledge, rather than data, is used. Data is raw information, not directly usable. As a consequence, it is not immediately obvious which pieces of data are appropriate to a specific application. Knowledge tends to be verbal rather than numeric; a piece of knowledge may be: "The machine is broken," whereas a piece of data would be: "$X = 5.25$." The use of knowledge rather than data makes the expert system easier to build, understand, update, adapt, and maintain.
- Knowledge and reasoning abilities are independent functions of the system. In a standard computer program, the procedures to be followed; the rules, case statements, and control algorithms; and the information concerning the task at hand are all bound together in one large system of programs and subprograms. To update the system, one needs to search through piles of control code to find the specific line of code that contains the information sought. In an expert system, the knowledge is stored in individual chunks (called *rules* and *facts*), without regard for the order in which they are to be used or the purposes they serve. The control and use of this knowledge are governed by an inference engine that has the ability to reason on any problem, as long as the necessary knowledge is made available.
- Decisions are based on heuristics rather than algorithms. A heuristic can be described as an imprecise procedure for solving a problem. Within this very imprecision lies the power of the heuristic approach: since a heuristic is loosely defined, it has a broader range of application and is a more robust approach. An algorithm will probably give a more correct answer to a problem than will a heuristic, but the heuristic will give a good enough answer far more often and will succeed in situations where the algorithm would break down and give no solution at all.
- The expert system emulates the thought process of a human expert. Humans tend to use heuristic approaches rather than algorithmic approaches, and their heuristics evolve over time. This is why a human expert is better than a novice at many tasks: he or she has built up a good system of heuristics to solve a variety of problems in a specific domain. Similarly, expert systems are most effective in tasks that require specific knowledge and procedures, rather than general reasoning.

From these concepts, we can begin to see some of the advantages of the expert system approach. First, expert systems are better than ordinary computer programs:

- Since they use heuristics, they are more robust and can find workable solutions to problems more often than can algorithmic programs.
- They are more easily updated, since knowledge is maintained separately from the reasoning mechanism. A new fact or rule can be inserted into the knowledge base easily, and the reasoning mechanism will use it when appropriate.

- They explain their solutions. As we shall see, expert systems easily can be endowed with a "justification" feature, so that they not only provide solutions to problems, but can explain their reasoning if asked.

Second, expert systems are better than human experts:

- They don't get sick, go on strike, or retire.
- They are easily duplicated. A human expert requires years of expensive training. To clone an expert system, the only cost is for an additional computer and a copy of the software (or for an additional terminal on a mainframe).
- Again, an expert system can explain its reasoning, whereas a human expert often has difficulty justifying decisions that he or she knows "just feel right."

B. Expert Systems vs. Artificial Intelligence

The concept of expert systems arose out of the artificial intelligence (AI) research community. It began as a tool for understanding the mechanisms of human cognition and for emulating human thought processes.

With the development of expert systems that had the ability to perform certain types of reasoning, it became clear that these symptoms could be used to perform useful tasks. A new segment of the AI community began to concentrate on developing expert systems to be used for their own ability, rather than as a research tool. New programming techniques were developed, and the expert system moved out of the realm of AI and into the world of computer science and programming technology, creating a new niche now known as *knowledge engineering*.

As with many other AI concepts, as soon as expert systems became useful tools with profitable applications, they ceased to be regarded as AI. Certainly, the development of specific expert systems for specific tasks is more knowledge engineering than AI work. But fundamentally, expert systems still conform to the goals and concepts of the AI community.

C. Expert System Structure

The structure of any expert system is based on four key elements, each of which has an analog in the human expert. The elements are the knowledge base, the inference engine, the user interface, and the explanation facility (see Figure 9).

The *knowledge base* is where the information resides in the expert system; it varies from one application to the next. Two expert systems, for two different purposes, may be identical in all respects except for the knowledge base.

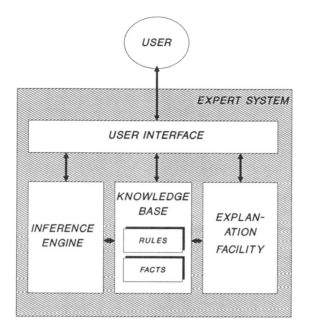

Figure 9 Expert System Elements

The knowledge base can be thought of as the "education" of the expert system. Its analog in a human expert is the years of training, experience, and practice that a person accumulates on the way to becoming an expert. Like the expertise of a human expert, the knowledge base is the most difficult element to acquire.

The knowledge base itself can be divided into two parts: the rule base and the fact base. The rule base consists of a collection of rules, often of the form "IF <condition> THEN <conclusion> " These rules are completely general, often include variables, and form the basis for decision making in all possible scenarios of the domain in which the system is an expert.

The fact base, by contrast, contains simple statements about the condition of the world for the current problem at hand. Generally, the fact base will change as the expert system is reasoning about a problem, whereas the rule base will change only if the knowledge engineer in charge decides that is should be updated.

Rules and facts together lead to new facts, in much the way that a syllogism functions in formal logic. For example, the rule "IF X is a man THEN X is mortal," combined with the fact "Socrates is a man," leads to the new fact "Socrates is mortal," which can then be added to the fact base.

The *inference engine* is the part of the expert system that performs the reasoning. It can be compared to the raw intelligence of a human expert. As with humans, different types of inference engines exhibit different levels of intelligence, but the same inference engine could be combined with many different knowledge bases to reason about many different domains of expertise.

Many different forms of inference engines exist, but all are designed to perform the same task: examine current facts and use available rules to generate new facts. One type of inference is called *discovery*. In this mode, all rules and all facts are used to generate all possible new facts, without regard to their possible relevance. In another type of inference, called *determination*, the use of the knowledge is more goal directed. A specific "hypothetical" fact is suggested, and only those rules and facts are used that are necessary to determine if the hypothetical fact is true or false.

The technique used for this reasoning is called *chaining* and both forward and backward chaining exist. Forward chaining is used for discovery (see Figure 10). Each fact in the knowledge base is compared against the IF part of each rule. If the conditions in the IF section are satisfied, the rule is "fired," which means that the inference engine will execute the THEN part, by creating the facts it specifies. These new facts are also compared to the IF portion of

---

Current Rules:	Current Facts:
1) IF A and B THEN C	1) A
2) IF U or V THEN D	2) B
3) IF C or D THEN E	3) Y
4) IF X or Y THEN F	4) V
5) IF E and F THEN G	
6) IF G THEN H	
7) IF G THEN I	
8) IF D and I THEN J	

Inference Engine Steps:
Rule 1 uses Fact 1 and Fact 2 to get new fact: C (Fact 5)
Rule 2 uses Fact 4 to get new fact: D (Fact 6)
Rule 3 uses Fact 5 to get new fact: E (Fact 7)
Rule 4 uses Fact 3 to get new fact: F (Fact 8)
Rule 5 uses Fact 7 and Fact 8 to get new fact: G (Fact 9)
Rule 6 uses Fact 9 to get new fact: H (Fact 10)
Rule 7 uses Fact 9 to get new fact: I (Fact 11)
Rule 8 uses Fact 6 and Fact 11 to get new fact: J (Fact 12)

Figure 10   *Discover* Using Forward Chaining

Current Rules: same eight rules as in Figure 10
Current Facts: same original four facts as in Figure 10

Task: Determine if J is a true fact.

Inference Engine Steps:
Goal is J
Rule 8 can prove J, so D and I are new goals
Rule 2 can prove D, so (U or V) is a new goal
V is a fact (Fact 4), so D is proven
I is only goal left to determine if J is true
Rule 7 can prove I, so G is a new goal
Rule 5 can prove G, so E and F are new goals
Rule 3 can prove E, so (C or D) is a new goal
D has already been proven, so E is proven
F is only goal left to determine if J is true
Rule 4 can prove F, so (X or Y) is a new goal
Y is a fact (Fact 3), so F is proven

All intermediate goals have been traced back to known facts, so the hypothesis, J, has been determined to be true.

Figure 11 *Determination* Using Backward Chaining

each rule, and so on, until no new firings are possible. At this point, all possible facts have been "discovered."

Backward chaining (see Figure 11) reverses this process, and it is used in determination. The expert system is asked to determine whether a candidate "fact" is true. The inference engine compares this fact to the THEN portion of each rule. If a rule is found that can generate the candidate fact, this rule's IF portion becomes the new candidate. The inference engine then proceeds to attempt to prove the validity of the facts in the rule's IF portion. Eventually, the hypothesis is either proven or the expert system runs out of facts and rules, thus determining that the original candidate fact is not true.

The *user interface* enables the expert system to communicate with a user. The form of the interface used depends on the intended audience of the expert system, and many systems contain a variety of user interfaces. The user may be well versed in the area of expertise in question and be able to respond to queries with little prompting or may be a complete novice in need of constant hand-holding. The user may be a computer expert who is comfortable with a crude interface or may be a neophyte who would appreciate complete user-friendliness with graphics, menus, and copious on-line help.

Whatever the form, the user interface must allow for the following communications between expert system and user:

- The user must tell the expert system what the task is.
- The expert system must tell the user what facts it requires.
- The user must tell the expert system the current state of the world as it relates to the situation in question.
- The expert system must convey its conclusions to the user.
- If asked, the expert system must be able to explain its decisions to the user.

The *explanation facility* is the element that allows the expert system to justify its conclusions. It is an important facility for several reasons:

- Users may not trust the expert system and may wish to be reassured that the conclusions are valid.
- Users may wish to learn from the expertise that the expert system possesses, rather than follow its suggestions blindly.
- The developer of the expert system may want to ensure that the knowledge base is properly codified.

The development of the explanation facility parallels the development of the rule base. Frequently, each rule in the knowledge base will include, in addition to an IF and a THEN clause, a WHY clause. This clause is used when the user requests an explanation of a conclusion.

D. Expert System Development

Now that a basic definition of an expert system has been established, we will discuss the hows and whys of applying expert system technology to aid in achieving TMA. Let us suppose that management has instructed us to develop an expert system to help make decisions in our inventory control area. We now face the task of going through the proper steps to construct a working expert system.

1. The Expert System Domain

The first task is to properly define our *domain*. In general, our domain for this job is inventory control. But we must be more specific. Will the job include keeping track of current inventory, or just making decisions? Will it calculate optimal order quantities? Will it allow for back orders? Will it provide answers to specific questions only, or will it volunteer information that the user may not think to request? We must decide what will be included before starting any development.

What defines an appropriate domain for an expert system? The domain must not be too broad. Handling the accounting for an entire corporation would be a Herculean task. A good rule of thumb is: Can one or two experienced people perform the task satisfactorily? If not, we may need to downsize our goals.

The domain must not be too general, either. Tasks that require "common sense" are rarely appropriate for expert system implementation. One reason for this is that even a human expert will have trouble explaining his or her reasons for making a decision. If something "just makes sense," we have very little basis for coding a rule in our knowledge base. A rule of thumb to keep in mind is that the task should be solvable with expertise that has been learned over the course of time—in other words, *experience*. If a novice is just as good at completing a task as an experienced expert, then we will probably have trouble developing an expert system to perform the task.

2. The Domain Expert

Once our domain has been specified and scoped appropriately, a *domain expert* is required. This is the source of the knowledge that will go into the knowledge base.

Frequently, the domain expert is an actual person, such as the person who has been performing the task in the past. If so, he or she may hestitate to participate in a project that will make him or her obsolete. While this has occasionally happened, a domain expert rarely becomes expendable after being immortalized in an expert system. For one thing, software requires as much maintenance as hardware; as company policy changes, the knowledge base must be updated. Furthermore, any large software project will require years of debugging and optimizing. In addition, the system will probably be implemented in several phases of gradually increasing complexity, and many years will pass before all tasks are automated.

As a final incentive to the balking human expert, there is the "mental wrench" effect. Often, the human domain expert who was used to generate the knowledge base becomes the user of the system in the future. The human expert finds that the expert system helps to magnify his or her own abilities within the domain. It frees him or her from the tedious parts of the job, allowing him or her to concentrate on the more creative aspects. The expert can set up complex "what if" scenarios to test out solutions to problems and can make decisions in a fraction of the time formerly required. He or she may even learn more about the domain through the experience of coding knowledge into a formal system.

3. The Knowledge Engineer

In the expert system development project, there must be someone who is well versed in the theory and mechanics of expert system technology. This person is referred to as a *knowledge engineer*. There is currently a great shortage of knowledge engineers in industry, and many companies are forced to hire consultants or attempt to create their own knowledge engineers from existing personnel.

The knowledge engineer is the bridge between the human expert and the expert system. While this person generally needs no prior knowledge of the domain area, he or she should be able to grasp and understand the principal concepts quickly. He or she must be enough of a psychologist to help the domain expert formalize his or her own knowledge, and must be enough of a computer scientist to code that knowledge into the form needed by the inference engine.

If knowledge engineering sounds like a daunting task, there is hope. With the development of newer and better shells (discussed below), the knowledge engineer's task is becoming easier. For many small-scale expert systems, the knowledge engineer and the domain expert are the same person. This is particularly true in technological domains, where the domain expert is already familiar with computing and programming techniques.

4. The Expert System Shell

In the "old days" of the 1960s and 1970s, the only expert system tools available were string-processing computer languages such as LISP and Prolog. These required large amounts of programming to accomplish even relatively minor tasks and ran on expensive mainframe computers.

Nowadays, a multitude of expert system "shells" exist on the market, with prices ranging from hundreds of thousands of dollars down to several hundred dollars. A shell takes care of the busywork that must be done to keep track of rules, facts, variables, and so forth. Shells are usually written in a standard computer language such as C or LISP and give the user a higher level "knowledge language" that is much easier to use.

The shell provides all elements of the expert system except the knowledge base. It includes an inference engine that can perform forward or backward chaining, or both; it includes some sort of ready-made user interface, often with a built-in menu structure and graphics interface; it includes an explanation facility that can communicate the "why" fields of rules to the user; and it often includes a "development environment" that provides basic tools for the knowledge engineer to aid in construction of the knowledge base. This could include a text editor, an interactive debugging facility, and sometimes an "intuition facility," which can extract rules from decision tables.

5. The Development Process

The very first step in the expert system development process, which must be completed before actual work begins, is the task of assembling the essential elements listed above. The domain of the system must be properly scoped, with a task that is well-defined, sufficiently focused, and capable of solution by the use of expertise. A domain expert, or experts, must be found; a person who can solve the task, a collection of well-documented case studies, or perhaps a good textbook. A knowledge engineer, or someone willing to learn to become one, must be hired or appointed to the task. Finally, the appropriate development platform must be acquired, probably an expert system shell with some sort of development environment.

a. Step 1—Structural Design. After these preliminaries are taken care of, the first step in the process is to design the structure of the expert system. The issues to address are such questions as: What will be the system inputs? What will be the outputs? What data is available? What kinds of rules will be used? What will be the principal variables? What will be the intermediate variables? Will there be different classes of rules? What will be the classes?

Answering these questions correctly is not essential. There will be opportunities to revise them. Most expert systems are developed organically, in that they grow gradually from a small prototype and evolve as they grow. It is best to make reasonable decisions on the structure of the system and then see how well they work before revising them, if necessary.

It is also wise to focus on a subset of the domain at this point and to code some rules in the chosen structure. If these rules seem to make sense, the structure can be considered usable.

b. Step 2—Knowledge Acquisition. At this point, a well-defined structure for the expert system knowledge has been developed. We have decided what sort of rules are needed, what they should do, and how they should interact. The next step is to extract the knowledge for the complete domain from the domain expert.

Attempting this step before the first step would be a mistake. The insights gained from the structural design phase help guide the knowledge acquisition process. This is especially important if the domain expert has trouble explaining his or her methodology. When presented with rule structures, the expert can usually "fill in the blanks" with less trouble.

Historically, knowledge acquisition is considered the bottleneck in the development process. It can be slow and tedious and can require many iterations and reinterviews with the domain expert. But take care that the step is performed completely and accurately. The quality of the finished expert system depends almost completely on the accuracy, precision, and completeness of the knowledge acquired.

c. Step 3—Coding. The next step is to convert the knowledge collected in Step 2 into the structure developed in Step 1. This should be fairly straightforward if the previous steps were handled conscientiously.

This is also the step in which the expert system shell's development environment comes in most handy. The task difficulty can vary by orders of magnitude between a well-organized development system and a poor one.

d. Step 4—Validation. After the knowledge is coded, it must be tested. Use the expert system from the perspective of the end user, assuming the level of understanding that he or she will have of the domain.

Run case studies, and compare the solutions offered by the system with those of the human expert. Initially, run the examples that were used to develop the rules in the first place. If these do not yield the expected solutions, there is a coding error. Once these examples have been validated, try some new examples that the system has not seen before. Compare its answers with what the domain expert would do in the same situation. In this way, the scope of the system can be determined. Ideally, the system should be able to solve as many types of problems as are likely to occur within the realm of its domain.

e. Step 5—Growth. As stated earlier, the best expert systems will grow organically, getting larger and more complete with time. Try following the first four steps on some significant subset of the knowledge domain first, and see how well the system works. If it performs adequately, add more and more capability. If not, do not be afraid to throw it away and start from scratch with a new structural design for the knowledge. The insight gained from the first attempt will make the second one that much better.

Also, be prepared never to put the project completely to bed. As time passes, new knowledge is created. The domain will evolve as product lines are updated, company policies change, and organizational goals are reevaluated. Occasionally, go through the knowledge base and weed out old rules that never fire anymore (some shells provide a mechanism for tracking rule usage). Proper attention to software maintenance will prevent the expert system from becoming just one more piece of code that used to serve a useful purpose but now collects electronic dust in some musty corner of your disk pack.

E. Expert System Topical Application Areas

Listed below are a variety of examples of expert system applications that have proven useful in various facets of the manufacturing industry. This is by no means a complete list, but merely some interesting topical applications.

1. Design
   a. Prediction of stress cracking
   b. Design of heat fins
   c. Evaluation of design for manufacturability
   d. Design of V-belt drive systems
   e. Design of wire ropes
   f. Design of reciprocating pumps
   g. Design of molds
   h. Cost estimation

2. Production
    a. Machining condition (speed, feed, etc.) selection
    b. Process planning for machining or assembly
    c. Job scheduling
    d. Inventory order control
    e. AGV path planning
    f. Labor resource planning
    g. MRP and MRP-III high-level control
    h. Robot selection
3. Control
    a. Fault diagnosis
    b. Alarm monitoring
    c. In-process inspection
    d. High-level control of PLCs
    e. On-line tool selection
    f. On-line tool monitoring
    g. On-line component failure prediction
4. Maintenance
    a. Troubleshooting diesel-electric locomotives
    b. Troubleshooting electrical systems
    c. Troubleshooting HVAC systems
    d. Troubleshooting industrial robots
    e. Diagnosis of vibration problems in turbomachinery

VI. SUMMARY

This chapter addressed the management control and leadership required as part of TMA. Control and leadership are fundamental and significant roles of management. If these tasks are accomplished effectively, the result is motivated and empowered people who provide innovation and excitement in support of the overall organization.

All departments must realize that they serve a common purpose and goal: corporate profitability and survival. All walls between departments must be eliminated and a broad focus on the manufacturing activity must be established by each. A free flow of information must exist throughout the corporation. Obviously, the corporate structure and mindset must facilitate this.

Organizational structure is important but the proper motivational environment must also exist. Depending on the situation, the appropriate leadership style must be used. However, leadership style flexibility must always be governed by the overall corporate motivational strategy developed and implemented.

Whenever we manage, a project management plan must be developed that defines and documents key project milestones and time frames. It is also productive to enhance a plan by defining a project management network or looking for the optimal solutions to decisions via mathematical routines.

Ideally, we always want to make accurate, timely, and consistent decisions for new (and, especially, repeat) situations. This is possible with the advent of expert systems. Expert system technology is widely available and enables the capture of human expert knowledge for continual reapplication simultaneously throughout the business and manufacturing environments of a corporation.

VII. QUESTIONS

1. Does TMA require an increase or decrease in management size? In management responsibility? In management integration? Why?
2. Identify several "walls" that exist in your organization. How might they be eliminated? Discuss several possible methods for each situation you can identify.
3. Chart the information flow existing for an environment that you are are exposed to (at work, at school).
4. What are the major factors involved in leadership?
5. Discuss the appropriateness of leadership flexibility.
6. What are the four major organizational behavior models? Discuss the characteristics of each.
7. Develop a management network diagram for a current work or school group project. Identify the critical path and the amount of slack.
8. Discuss the trade-off between constrained optimization model complexity and accuracy.
9. What does "feasible region" mean?
10. Discuss the difference between "knowledge" and "data." Give three examples of each.
11. Discuss the difference between an "algorithm" and a "heuristic." Give two examples of each.

VIII. REFERENCES

1. Bowers, D. and Seashore, S.: Predicting Organizational Effectiveness with Four-Factor Theory of Leadership. *Administrative Science Quarterly*. 238-263, Vol. 11, 1966.
2. Tannebaum, R. and Schmidt, W.: How to Choose a Leadership Pattern. *Harvard Business Review*. 95-101, Vol. 36, 1958.
3. Vroom, V.: Decision Making and the Leadership Process. *Contemporary Business*. 47-64, Vol. 3, 1974.

4. Davis, K.: Evolving Models of Organizational Behavior. *Organizational Behavior*. McGraw-Hill, New York, 1985.
5. Vroom, V.: *Work and Motivation*. John Wiley and Sons, New York, 1964.
6. Deci, E. and Gilmer, B.: *Industrial and Organizational Psychology*. McGraw-Hill, New York, 1977.
7. Lawler, E. and Porter, L.: Antecedents of Effective Managerial Performance. *Organizational Behavior and Human Performance*. 122-142, New York, 1967.
8. McClelland, D.: *The Achieving Society*. Van Nostrand, Princeton, New Jersey, 1961.
9. Maslow, A.: *Motivation and Personality*. Harper and Row, New York, 1970.
10. Vroom, V. and Deci, E.: *Management and Motivation*. Penguin, Baltimore, 1970.
11. Davis, K.: *Organizational Behavior*. McGraw-Hill, New York, 1985.
12. Kelly, A.: *Maintenance Planning and Control*. Butterworths, London, 1987.
13. Moore, F.: *Manufacturing Management*. Richard D. Irwin, Inc., Homewood, Illinois, 1969.
14. Eppen, G. and Gould, F.: *Quantitative Concepts for Management: Decision Making Without Algorithms*. Prentice-Hall, Engelwood Cliffs, New Jersey, 1985.
15. Parsaye, K. and Chignell, M.: *Expert Systems for Experts*. John Wiley and Sons, New York, 1988.
16. Waterman, D.A.: *A Guide to Expert Systems*. Addison-Wesley Publishing Company, Reading, MA., 1986.
17. Patterson, D.W.: *Introduction to Artificial Intelligence and Expert Systems*. Prentice-Hall, Englewood Cliffs, New Jersey, 1990.
18. Kusiak, A.: *Intelligent Manufacturing Systems*. Prentice-Hall, Englewood Cliffs, New Jersey, 1990.

Part Three

MANUFACTURING SYSTEM CONTROL

Men are born with two eyes, but with one tongue, in order that they should see twice as much as they say.

—Colton

5

System Definition

All manufacturing systems must be designed with robustness and efficiency in mind. This means that the system must fulfill a set of comprehensive and detailed state-of-the-art design requirements. It must also achieve this in a manner that is not overly complex and cumbersome.

This chapter addresses key topical areas that are considered when defining an efficient and robust manufacturing system. These areas are addressed in four sections entitled plant layout, material handling, automation, and simulation.

I. INTRODUCTION

In any major undertaking, no great structure is built without first laying a firm foundation. For our task of attaining TMA, this translates to the solid base of a well-defined operating environment. Labor, machines, and materials, as well as information and instructions, must be organized efficiently, according to the best standard industrial engineering techniques. Without a sound basic structure, will all parts smoothly interacting, striving toward TMA may become moot.

This chapter examines the basics of manufacturing system definition. This includes looking at basic management and organizational functions typical of a modern industrial engineering curriculum, plus innovations that industry is currently adopting. While exploring this chapter, keep in mind the hierarchy involved in the manufacturing-based corporation: manufacturing system, subsystem, equipment, module, part.

II. PLANT LAYOUT

A system is fundamentally composed of various equipment. This may include equipment for machining, assembly, and material handling, as well as storage, loading docks, repair facilities, computers, and personnel offices. Where are these to be located? Plant layout or, in broader terms, *facilities design*, enables us to answer this question using analytical and design techniques aimed at maximizing efficient organization and interaction between various assets.

We are looking for more than a floor plan. There must also be an adequate power supply; lines of communication for both voice and data; and, most importantly, room to grow and adapt to inevitable changes in the future.

Many techniques used in facility planning involve minimizing necessary evils: for example, minimizing the amount of equipment needed to facilitate economical production, decreasing distances to be traveled by material in-process to save time and energy, and lowering the inventory cost of material tied up in production. A layout that allows fewer workers to attend to a greater number of tasks results in labor savings. A sample list of items to minimize includes:

1. Investment in equipment
2. Production time
3. Material handling
4. Work-in-process
5. Floor space
6. Variation in equipment
7. Inflexibility
8. Labor
9. Employee inconvenience, discomfort, and danger

A. Plant Layout Analysis

Now that we know the major goal(s) of plant layout design, we need to address the fundamental question: How do we arrive at the layout that achieves those goals? Many techniques exist, both new and old, which all follow one of two basic schemes:

1. Select a number of candidate layouts. *Analyze* each layout according to some grading method. Select the layout with the highest score.

2. Define a list of goals, constraints, and requirements for the target layout. *Synthesize* the best possible layout using some optimization technique.

These two basic schemes are referred to as the *analytic* and *synthetic* methods of plant layout. We shall begin by discussing some of the analytic techniques available.

1. The From-To Chart

The From-To chart, also called a travel chart, is a matrix that tracks flow between the various locations in a plant [1]. Each activity location (machine, storage center, entry or exit point, etc.) is listed across the top row of the chart and down the first column of the chart. Each element in the interior of the chart lists some fact concerning the relationship between the corresponding row and column locations (see Figure 1).

One use of a From-To chart is to list distances. An element in row 4 and column 2 would show the distance *from* location 4 *to* location 2. Naturally, the diagonal would have all zeros, and the chart would be symmetrical about this diagonal.

Another From-To chart could be used to tabulate traffic volume. The numbers in the interior of the chart would indicate the flow of parts from each

TO ⟍ FROM	Receiving	Storage	Lathe 1	Lathe 2	Vert. Mill	Horiz. Mill	Packaging	Inspection	Shipping
Receiving		40		2		5			
Storage			8	3	10	12	5		10
Lathe 1				9	22	13	6	7	
Lathe 2			7		3	11	15	9	
Vert. Mill	5	11	6			23	6	1	
Horiz. Mill			5	4	17		11	21	
Packaging		8						10	30
Inspection		6					11		27
Shipping									

Figure 1   From-To Chart for Part Flow Between Departments

location to each other location. In this application, of course, there would not necessarily be diagonal symmetry.

Combining these two charts, the distance From-To and the traffic From-To, gives the total material handling burden of the proposed layout. Slight changes in the layout can be documented easily by altering one row and one column. In this way, many candidate layouts can be compared.

2. The REL Chart

Another analytic tool is the REL, or *activity relation*, chart [2]. This technique is more subjective than the From-To chart and is useful for considering the more qualitative influences, as well as distance and traffic volume. The REL chart uses a *closeness rating* instead of a numerical score for each location pair (see Figure 2).

Unlike the From-To chart, which listed facts about a specific layout, the REL chart is a method for expressing desired distance relationships. The closeness rating for each pair of locations is filled in based on a qualitative judgment of how important it is that the two locations be situated near, or even adjacent to, each other. The standard ratings are

A = Absolutely necessary
E = Especially important
I = Important

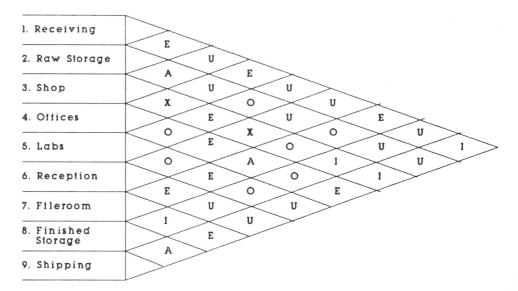

Figure 2   Activity Relation (REL) Chart

O = Ordinary closeness OK
U = Unimportant
X = Undesirable

These ratings are selected intuitively, based on the subjective judgement of the plant layout engineer. Often, a reason code is listed as well, indicating the rationale for the closeness rating. Typical reason codes might be sequence of work flow; sharing of records, personnel, or equipment; similarity of function; or ease of supervision. Undesirable ratings may arise between noisy operations and those that need quiet surroundings, or particularly dirty or sooty operations and those that require a clean environment.

3. Space-Relationship Diagram and Block Plan

We are now ready to start constructing some candidate layouts and evaluating them. The REL chart is our starting point for generating the candidates, and the From-To chart will enable us to judge them.

The space-relationship diagram is the first step. This diagram combines two sources of information: our REL chart and our space requirements. For each activity listed in the REL chart, we must calculate the necessary amount of floor space. This can be derived from anticipated production volume, time standards, and machine capacities.

Next, a square is drawn on a piece of paper for each activity. The size of each square is proportional to the area needed for that activity, and this area is listed in the square, along with the activity number (see Figure 3). Each square is connected to each of the other squares with a line.

The type of line used is based on the closeness rating assigned in the REL chart: An A pair is connected by a quadruple line, an E pair by a triple line, an I pair by a double line, an O pair by a single line, and a U pair by no line at all. Activity pairs with an X rating are connected by a zigzag line.

This space-relationship diagram is the tool for generating the candidate layouts. Think of the lines connecting each activity as attractive forces. The more lines between two blocks, the stronger the attraction, like so many tightly stretched rubber bands. The zigzag lines, of course, must be thought of as repulsive forces.

The next step is to try to assemble the activities into a block plan. Using your imagination, let the attractive forces bring the activities as close together as the relationship lines indicate. Think of the process as piecing together a jigsaw puzzle with rubber bands connecting the pieces. The final shape should approximate the shape of the plant or building that will house the activities. A typical block plan is shown in Figure 4.

Naturally, many block plans are possible from any one given space-relationship diagram. Try to generate as many reasonable plans as possible. The

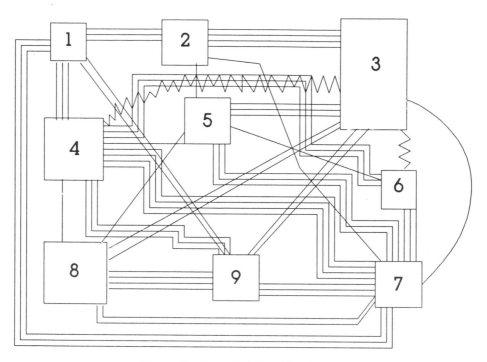

Figure 3   Space-Relationship Diagram

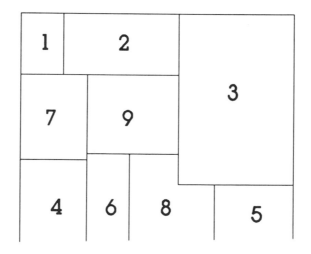

Figure 4   Example Block Plan

more candidates that are generated, the greater the probability that the best possible plan is among them.

When you have sufficient candidate plans, the From-To chart can be used to evaluate them. The traffic volume From-To chart will be the same for each candidate, but the distance chart will not. Measure the distances for each pair of activities in each candidate and construct the appropriate distance chart. By combining this with the (constant) traffic volume chart, each candidate can be given a score for material handling burden.

A simple final step would be merely to select the candidate with the lowest material handling score, but this can sometimes lead to undesirable results. It is better to select the best three or four candidates and scrutinize each closely. Are the closeness relationships implemented as desired? Are the A pairs and X pairs situated as desired? It is possible, especially in complex systems, for one or two important relationships to be sacrificed in favor of many small improvements in the less important relationships. This results in a better numerical score, but lets a weakness slip into the design. Use sound engineering judgment to select the final winner from the top three or four contenders, based on overall adherence to closeness ratings and any other subjective considerations not programmed into the rating scheme.

B. Plant Layout Synthesis

The preceding subsection focused on *analysis* of candidate layouts, with only minimal effort directed at the generation of those candidates. This is a tried and true method of plant layout and works well for small plants with few departments. With large, complex facilities, however, the method of creating these candidates becomes more important.

Synthesis of candidate layouts requires many complex decisions to be made as each function or activity is placed into a layout. Consequently, computer methods are often used. Many programs for plant layout exist, three of the most popular being CORELAP [4], ALDEP [5], and CRAFT [6]. Some programs employ a *generative* approach, building up each candidate layout from scratch, while other use a *derivative* approach, which continually modifies an existing layout, making incremental improvements at each step. Some programs are capable of using either approach or a combination of the two.

Although different programs use different methods of generating layouts, they all have certain functions in common. Most of them require as input the REL chart, as described in the preceding section. They also require a *grading scale*, which translates the codes in the REL chart to a score for each candidate layout. Some programs assume a constant grading scale, others allow the user to supply one.

The programs must also be told how much area is required for each department and what is the size and shape of the available facility. Some programs can handle multiple floors at one time, and some allow the user to fix certain activities or departments in specific locations before synthesis begins.

Most generative programs begin by selecting one activity from the REL chart. The method of selecting the first activity varies from one program to another, but is usually based on some function of the REL chart that allows the greatest flexibility in subsequent selections. This activity is placed in the layout, and the REL chart is searched again for an appropriate second activity. Again, each program uses its own criteria for selection of the order of placement, based on some function of the REL chart and the grading scale. Each successive activity is placed in the layout at a location that will minimize or maximize some program-specific function. The program generates a large number of possible layouts, giving each a score based on the grading scale.

The best of these first-cut layouts is stored as a benchmark, and the program then attempts to do better. Some programs will start over with a different first activity. Others will go into a derivative mode and attempt to make improvements on the benchmark, using various internal functions.

Finally, after several iterations, the program will report its latest and greatest layout. One then must decide if this layout meets all the nonnumeric requirements that were not programmed into the analysis. Generally, one also has the option to supply a different grading scale or a modified REL chart to see if a preferable layout results. As always, one should take into account past layout experience and engineering intuition in making the final decision, and not take any computer program's recommendation as gospel.

III. MATERIAL HANDLING

In the previous section, we saw some techniques for laying out the facilities in a manufacturing plant. These techniques were based on several principles, one of the most important being minimization of the material handling burden.

Is material handling cost worth worrying about? Absolutely—especially in a TMA environment. For one thing, material handling cost can be up to or even beyond one-half of total manufacturing cost. Furthermore, in many applications, material handling is inextricably linked with other manufacturing elements. Conveyors and transfer mechanisms in assembly lines, part feeders and orienters, and robotic manipulation equipment are examples of devices that handle raw material and work in process at the same time that they are being converted to finished products.

What materials are handled in a manufacturing system? Basically, everything from raw materials to work in process to finished products. Also tools, replacement parts, energy, information, and sometimes even personnel are moved by the material handling systems.

What are our goals in handling these materials? First of all, we want them to reach the designated destination, at the proper time or earlier. They must arrive safe and undamaged, and in an economical manner. Further, we should always be aware of each item's location while it is in transit, in case a change in plans makes it necessary to reroute.

Material handling equipment covers a broad spectrum of sophistication, from trivial to highly automated and intelligent. In the most basic sense, labor can be considered a material handling technique: workers can trot over to a storage area and retrieve the materials they need. This material handling device has the lowest investment cost, but one of the highest unit costs. As such, it is advisable only for very low production rates.

Next, we have human-powered equipment such as dollies and hand carts. Then come human-controlled but autopowered devices such as fork lifts and motorized carts. Hoists, gantries, and conveyors require human control only for starting and stopping at desired times. Finally, we have intelligent and automated devices such as AGVs (automated guided vehicles) and conveyors with feedback control.

The best material handling method for each application depends on a combination of factors. Quantity of material to be handled is obviously an important consideration, from the standpoints both of how many units per day and of how many days in the production run. High material flow rates require large capacity systems, and longer production runs justify a larger investment in faster and more sophisticated equipment.

Another critical factor is flexibility. Will the handling system remain constant for a long time? Will rerouting become necessary often? Are changes in flow rate expected? Naturally, greater flexibility comes at a greater cost and must be justified both by anticipated changes in material handling requirements and by enhancement of profitability resulting from the flexibility.

Certain guidelines should be followed when selecting and designing the handling system. These concepts are basically common sense, but are still worth discussing briefly. The first guideline is the *bee-line principle*: whenever possible, use straight lines to minimize distance traveled. Of course, practical considerations may overrule this concept in many situations.

Next is the *pallet principle*: unless we are dealing with extremely large items (say, locomotives), it is a waste to plop one unit onto a conveyor. This wastes conveyor capacity, as well as loading and unloading motions. It is better to palletize parts onto a single holder, or pallet, and manipulate them as a single package. The pallets should be as large as possible and should be of consistent size, arrangement, and orientation.

Along with the pallet principle comes the *buffer principle*: build buffers into the production line so that full pallet loads can accumulate before transportation becomes necessary. This not only makes palletizing more efficient, but also allows different workstations to work at slightly different rates without blocking the flow lines.

Another guideline is the *minimum dead-time principle*: we want to keep the material moving, not sitting around. Try to speed up the loading and unloading of parts onto and off of the handling equipment. This means not only well-designed pallets and fixtures, but also intelligent scheduling of loading and unloading operations during our overall process.

The *two-way principle* is easily overlooked, but can double handling efficiency: never send anything back empty if you can avoid it. Use the same pallet that arrived at station n to take station n's output to station $n + 1$. Don't send it back to station $n + 1$ for another load! Arrange flow so that minimal amounts of motion are wasted. Here is where uniform pallet design pays off.

Finally, we have the *information principle*: whenever possible, integrate the information about the material into the flow with the material. Rather than saving a list of part numbers and their associated pallet numbers in a computer file for later retrieval by a downstream station, print the part numbers onto the parts or pallet itself. Bar code printers and readers, or even less sophisticated methods for low volume, make this possible.

A. Material Handling Equations

The relationships describing various quantities in material handling systems are fairly simple. One basic concept is the rate of flow, R_f usually expressed in parts per minute. For a one-way conveyor or other handling system, it is given by

$$R_f = n_p \frac{V_c}{S_c} \leqslant \frac{n_p}{T_L}$$

where n_p is the number of parts per pallet, V_c is the flow velocity (feet per minute), S_c is the spacing between pallets (feet), and T_L is the loading time at each end (minutes).

A material handling system also has a capacity, or number of parts that can be accumulated in it. This can be thought of as a temporary storage of work-in-process. Letting n_c be the number of carriers (pallets) in the system, then

$$n_c = \frac{L_f + L_r}{S_c}$$

where L_f is the length of the forward (full) part of the route and L_r is the length of the return (empty) part of the route.

Including a return portion L_r makes this equation applicable to recirculating types of systems, such as belt conveyors. The total number of parts in the system, N, is given by

$$N = n_p \frac{L_f}{S_c} = n_p n_c \frac{L_f}{L_f + L_r}$$

One other important relationship should be noted here: the relationship between travel time and loading/unloading time. If T_L is time to load a pallet and T_u is time to unload a pallet, then

$$\frac{V_c}{S_c} \text{ must be } \geqslant \frac{1}{T_L} \quad \text{and} \quad T_L \text{ must be } \geqslant T_u$$

to avoid backups in the system.

Note that other material handling relationships, both specific to individual systems and applicable in general to all types, are found in the literature [8].

B. Automatic Guided Vehicles

The highest end of the material handling spectrum is occupied by AGVs. These devices are by far the most expensive and therefore are not applicable to all situations. Their vast capabilities and inherent flexibility, however, can lead to large payoffs for installations that can justify their cost.

AGVs are basically independent vehicles, rolling from station to station under their own power and control. Some are guided by wires or painted lines on the floor, others by dead reckoning with periodic alignment checks at known landmarks. They are generally powered by on-board batteries that last at least an eight-hour shift before needing recharging. Often they have a "safety bumper," a large loop of some flexible material on the front surface, which will detect collisions, and stop the vehicle, before the main mass of the AGV does any damage to itself or other objects in its way. Some AGVs use optical or sonar devices for collision detection.

Different control schemes can be used with AGVs. In some systems, the AGV is controlled by typed instructions entered at an on-board control panel. This is a simple and straightforward method, but requires large amounts of human intervention. In other systems, a decentralized control scheme is used: a station in need of an AGV will issue a general service call on a radio frequency. Any AGV with unused load capacity can detect and answer the call and service the station. This is a useful scheme in large factories with "islands of automation," but no centralized control of the entire floor. If a central computer is used to direct flow in the entire factory, it will generally make the decisions as to which AGV should service which stations, and when. This method offers the greatest overall efficiency, but requires extremely high levels of planning and programming.

Equations to plan an AGV system are similar to those given above for any material handling system, with the addition of a traffic factor to account for time wasted in waiting for obstacles to pass, avoiding collisions, adjusting a path, and so forth. This factor, usually somewhere between 0.8 and 1.0, is used to reduce the number of parts per minute that could be handled by an AGV with no need to worry about traffic congestion. Naturally, the more AGV and other traffic in the system, the lower the traffic factor will become.

IV. AUTOMATION

Automation, from the Greek words for *self* and *moving*, is one of the most common buzzwords in manufacturing today. This section looks at some of the basic concepts of automation and examines some of its strengths, limitations, and trade-offs.

The word automation can be applied to all parts of the manufacturing enterprise where direct human control has been minimized or even eliminated. This includes material handling, machining, assembly, inspection, storage and retrieval, and any other tasks that are carried out in the process of converting raw materials into a useful and salable product.

Automation is almost universally accepted as a good thing. The reasons for this acceptance are compelling. Productivity increases are an obvious advantage of automation, in terms of both production per hour and production per worker. Other advantages of automation are decreased unit cost, increased safety, increased product quality, reduction of work-in-process, and greater predictability of the manufacturing function.

Basically, automation can be broken down into two broad classifications: hard and soft [9]. Hard automation, the older type is based on machines that do only one thing, but do it extremely well. The automobile industry was dominated by hard automation a decade ago and still makes extensive use of it. Examples of hard automation include automatic presses and die-casting machines, automated transfer lines, and any other heavy machinery that performs the same task over and over. It is characterized by a lack of flexibility, long production runs, and high production rates.

Soft automation, on the other hand, is characterized by machines that can perform a variety of tasks. Soft automation is also called *flexible* or *programmable automation* and is dominated by robots and other general-purpose, programmable devices. The idea behind flexible automation is that a single investment in equipment will pay dividends in the production of many different products and processes. One cost of this flexibility is that the tasks are sometimes performed at a lower production rate than if dedicated, single-purpose machines were used.

Which type of automation is best for a given application? The answer to this question depends on many factors, including the type of industry, the anticipated production rate and run length, frequency of product changes, and future business plan. For any one given manufacturing task, it is much cheaper to acquire hard automation equipment that can perform the task than to use soft automation equipment.

However, if the task becomes unnecessary, the hard automation equipment will require extensive modification, if not outright scrapping. The flexible equipment can be modified easily for a new task via software alone, with no further investment in hardware. Furthermore, the flexible equipment can be instructed to perform the new task along with the old task, as each task is needed, if a variable product mix is desired.

Finally, it should be noted that the distinction between hard and soft automation is not as clear-cut as it is often made out to be. In truth, a sort of continuum exists, wherein any level of flexibility that is desired can be achieved. There are individual devices that exhibit certain degrees of flexibility, and there are entire factories that contain elements of the programmable along with the inflexible.

The amount of flexibility that is selected for any specific installation should be based on the amount that is needed and anticipated. To pay for excess flexibility that will never be used would be an unwise waste of funds.

A. Automation Elements

All automation is based on certain types of equipment, capable of performing standard automation tasks such as material handling, assembly, inspection, and so on. Each type of equipment, with the exception of robotic devices, is available in varying degrees of programmability, and so should be considered key elements of both hard and soft automation systems. Robots and similar devices, of course, are inherently flexible and should be considered elements of soft automation only.

Transfer mechanisms move a workpiece from one station to the next in an automated fashion. Several configurations exist. An in-line transfer mechanism consists of a conveyor or overhead chain system to move the workpiece along in a more or less straight line; a rotary mechanism is more like a turning table, around which the necessary processing stations are arranged like guests at a dinner table. The rotary configuration is useful only for situations with a small number of workstations, whereas an in-line system can accommodate as many stations as are desired merely be increasing its length. Both types can move in either a continuous or indexed fashion, depending on the types of operations to be performed at each station.

Automated machining devices perform the same functions as traditional machine tools, but are directed automatically with some sort of computer control, generally CNC or DNC (direct computer control) (discussed below). These machine tools can consist of traditional tools such as drill presses, vertical or horizontal milling machines, lathes, presses, and the like, or more modern devices that combine several of these functions. The machines may operate on a stationary workpiece positioned by the transfer mechanism, or may have a built-in multiaxis positioning table to hold the work.

With regard to the newer multipurpose machine tools, many types of configurations are available. These are often called *machining centers* and may have several spindles that are driven by a common motor and are capable of coming to play on the workpiece in whatever order is required by the process plan. A typical machining center may combine a drill press, vertical and horizontal mills, and a tool holder.

PLC, CNC, and DNC refer to methods of using computers to control automated machine tools. They are discussed here in order of increasing sophistication. However, keep in mind that new hardware and techniques are constantly being developed, with the result that clearly drawn distinctions among these three types of control are becoming more difficult to make.

PLC stands for programmable logic controller, the original device used to control automated machine tools. A PLC is a dedicated device attached to one tool and often is supplied by the machine tool manufacturer as a part of the tool system. It is basically a digital computer that can store a sequence of operations for the tool and send the appropriate commands to the tool at the appropriate times. The input console of a PLC often consists of a collection of buttons corresponding to the standard commands for the tool. A small screen is often included so that the program can be reviewed and edited.

The PLC bears little resemblance to a general-purpose computer, but is more of an intelligent command console with a memory. Some are equipped with disk drives so that programs may be saved, and some have serial ports allowing them to download programs from some central computer. PLCs have been popular since the late 1960s and continue to be important industrial components today.

The DNC concept moves toward the use of general-purpose computers, rather than specially designed digital devices, to control machining processes. In DNC, or direct numerical control, a central computer stores the programs for a large number of machines. A DNC computer can be thought of as one large PLC, managing all the machines in a factory or part of a factory. This allows for coordination of machines and processes, as well as monitoring of overall factory activities.

CNC, or computer numerical control, was a natural outgrowth of DNC that was made possible by the decrease in size and increase in power of general-purpose computers. A CNC computer controls only one machine, or possibly a small number of machines working together. The advantages of CNC over DNC arise from the localization of control. An engineer on the factory floor has direct access to the computer at the site of the tool. Also, since CNC computers control a single machine, there is no need for time sharing or other resource allocations, and the complete power of the processor can be used in complex control algorithms.

There is no danger of a single computer failure crippling the entire factory, as in the DNC implementation. CNC computers are usually networked to a central computer, so that there is no loss of the DNC ability to monitor all processes and keep all elements of the factory in communication with each other element.

Robots, or more precisely, programmable manipulators, are the final element of manufacturing and help to pull everything together. Robots have many applications in an automated factory, including loading and unloading pallets, applying sealants and adhesives, cutting, inspection, and assembly. Perhaps the most popular current uses of robots are painting and welding. Any task that involves the manipulation of some material or device through a precisely defined trajectory can be accomplished with some type of robotic device.

As with any programmable automation element, the high price of a robot must be justified through savings in manufacturing cost. Tasks that are performed many times per hour, perhaps for two or three shifts per day, are especially suitable for robots. Also, tasks performed under hazardous conditions, requiring many expensive protections and precautions for human operation, may be done much more cheaply by a robot.

Several issues must be kept in mind when considering a robot for a task, based on the robot's capabilities and the task requirements. Chief among these is accuracy: Can the robot position and orient the object to be manipulated with the required accuracy? Related to accuracy is repeatability. Repeatability is often defined as a robot's ability to reach the same location several times in a row, without drifting as time goes by. Another issue is resolution, the smallest increment of motion possible for the robot. Is the resolution small enough to perform the most delicate of the required motions?

Other considerations include the load-carrying capacity, the reach of the arm, and the type of grippers (also called *end effectors*) available for the robot. Another factor is the capability of the robots controller: how programmable it is, and what kind of sensors it can be connected to. These capabilities vary greatly from one type of robot to another, and a wide variety of options are available on the market today.

V. SIMULATION

Assume that we have developed a tentative layout of our new manufacturing system, proposed changes to our old system, or perhaps just a new control scheme for a physically unchanged system. Before implementing our new ideas, which could cost millions of dollars, it would be nice to have some evidence that our ideas are sound. Ideally, we would like to conduct a scale-model experiment, a dry run of the new plan, before we commit ourselves.

That is exactly what simulation allows us to do. General-purpose computer-based simulation is a programming technique that allows us to analyze systems with random variables in a manner that gives us fairly good estimates of the system's performance.

The concept of random variables is crucial. In most complex systems, we cannot predict the future exactly. Sales volumes, consumer demand; breakdown times, frequency, and severity; randomly varying production times; and many other system variables are not known precisely beforehand. If they were known, we could develop mathematical relationships between these inputs and our system outputs, although these relationships might be extremely complex.

However, with simulation techniques, we do not need the exact inputs. All we need is a mathematical expression for the distribution of the inputs. Typical distributions are

1. Normal—with known mean and standard deviation
2. Exponential—with known mean
3. Poisson—with known mean
4. Triangular—with known minimum, maximum, and mode
5. Uniform—with known minimum and maximum

These distributions are described by fairly simple mathematical expressions and can be derived from historical data of the process in question. As long as no change is expected in the overall distribution of a random variable, the distribution developed from the historical data can be considered an excellent prediction, over time, of the future events. These predictions become the inputs to our simulation.

How do we develop our simulation? First and foremost, we must define our system precisely. That is, we must decide exactly what parts of our factory we wish to model. We must draw a "dotted line" around the part of the factory that will be included in the model. We must scrutinize this dotted line and determine all inputs and outputs to the system. These include everything that crosses the dotted line, either in or out. We must identify material, personnel, products, energy, and information that travels in and out. Anything that happens completely inside the dotted line will become part of the internal workings of the model; those things that cross the line are modeled as inputs and outputs.

What exactly is a model? In a general sense, a model is an abstract description of some real system. It is an analogy whose behavior can be used to infer the behavior of the system it models. Many types of models exist in the world, some of which are used in computer simulation. Scale models, such as miniature trains, cars, or soldiers, are useful to architects and generals. Graphical models are useful in other situations; a classic example of a graphical model is a road map.

The two model types we will be using are math models and logical models. An example of a math model is an equation, such as $F = ma$. This model describes the behavior of the quantities of force, mass, and acceleration and can be used to predict future events, if the inputs are supplied correctly. Another example of a math model is a look-up table: given row and column values as inputs, the output can be looked up in the interior of a table of values. This type of model is most applicable to functions that cannot be described by a simple equation.

A logical model is a description of inputs and outputs in terms of logical statements. For example, a simple IF-THEN rule could constitute a logical model. A truth table would also fit the definition of a logical model. A complex FORTRAN program, with many branching IF statements, is a large logical model of some decision process. An expert system (see Chapter 4) is another example.

Our system model will include both math and logical models in its attempt to mimic system behavior. It will include those portions of the system that we consider important and will describe them to a level of detail, or resolution, that we consider necessary. It will be supplied with the quantities we identify as system inputs and will return to us the system outputs.

When the model has been completed, it will enable us to do four things that we could not have accomplished as easily, if at all, before:

1. If the model can be made to perform exactly as the current system does, it can be used to EXPLAIN the real system. The math and logical models we select must be actually functioning in the real system.
2. The model can PREDICT how the system would react to a new set of inputs or to a change in the system itself.
3. This allows us to ANALYZE the system and its performance.
4. Iterative analysis of progressive changes to the system model allow us to DESIGN a new system, or changes to the old system, in accordance with our new goals.

VI. SIMULATION EXAMPLE

What does a typical simulation look like? We can examine a simple example that contains many of the elements of a more complex situation. We will look at this model merely in logical terms and will not worry about programming or implementation.

A typical system with random variables can be demonstrated by the operation of a fast-food restaurant. Let's assume that our restaurant has one server and that customers wait in one line to be served. This line is usually called a *queue*.

First, we must define the state of our system. The *state* is a set of variables that completely, yet without redundancy, describes the condition of the system. Any variable not included in the state description can be derived from those that are included. Many possible state descriptions exist for any given system. For our fast-food system, the state can be described by two variables: the server status (either busy or idle) and the number of customers waiting in the queue, not counting the customer, if any, being served by the server. We will assume that the server works diligently at serving any waiting customers and becomes idle only when the queue is empty.

Next, we must identify the ways in which the state can change. In our example, there are two ways: a new customer can arrive, or the server can finish serving a customer, who then leaves. Identifying these state-changing activities is something of an art and requires a bit of practice.

Next we examine how these activities change the state. Generally, the manner in which the state changes depends not only on the activity that occurs, but also on the state of the system just before it occurs. Our first state-changing activity is a new customer arriving. If the server is idle, the new customer gets immediate service. This means he or she does not get in the queue, so the queue length stays at zero. The server status, however, changes to busy. If the server had been busy when the new customer arrived, the customer would have to wait, increasing the queue length, but not affecting the server status.

The other activity is the server completing service of a customer. If there are no customers in the queue, server status changes to idle. If there are customers in the queue, server status remains at busy, and queue length decreases by one.

Now we have completely defined our system and identified how it will function. To set it in motion, we must supply it with some inputs. The inputs will be the arrival times of each customer and the time required for service by each customer. Each of these values will be supplied from some random distribution, supposedly derived from a study of past data concerning customer interarrival time and service time.

Output statistics are completely up to us. We can program our simulation to collect whatever statistics we are interested in, as long as they are based on variables in the model. For our example, we would probably be interested in things like average time spent in the queue by the customers, average queue length, and percentage of the time that the server is idle.

Now we are ready to let the simulation begin. We give it an initial state (usually server = idle and queue = empty) and let the program run. The program selects a customer arrival time from the appropriate random distribution, as well as a service time. It selects the next arrival time, and if this is sooner than the first service is completed, it increased the queue length by one. As each customer arrives and leaves, the state is noted and statistics are recorded. After sufficient simulation time has passed, we will have a full set of output statistics on the values that we requested. These outputs, of course, will be statistical in nature, mostly averages and standard deviations. But there is no way we could have merely calculated them from average input statistics; the relationships, in general, will have been far too complex.

What about the actual coding of this model into a computer program? Even for the relatively simple example of the single-server/single-queue system given above, we could be faced with a large programming task. Fortunately, commercial packages are available that make this task fairly simple. Most systems are composed of basic building blocks, which tend to look the same, from the viewpoint of a model, regardless of the physical system they are based on. Simulation packages supply ready-made program chunks for each of these building blocks. The user needs only to define the blocks he or she needs, define the interconnections between them, and supply actual values for the variables.

Example of these building blocks, also known as *network elements* are

Arrival elements—model customers, orders, workers, and soon, as they enter the system. These elements must be supplied with interarrival times for the arrivals.

Queue elements—model a queue in the system. These elements must be supplied with an initial queue length and, possibly, with a maximum allowable queue length.

Activity elements—model an activity that takes a certain amount of time. These elements must be supplied with a duration and with how many customers can be served at the same time.

Conditional elements—allow decisions to be made in the system, based on system state.

Statistical elements—allow the program to collect statistics based on current model state.

Departure elements—model the departure of customers, orders, and so on, from the system.

Other, more complex elements also exist, but these tend to vary from one simulation package to another. Each has its own strengths and weaknesses. Some are more geared toward specific situations or industries than others. Some have built-in animation and graphics packages, which can display a schematic diagram of the system model on a computer screen as it operates, allowing the user to watch customers arrive, get service, wait in queues, and so forth.

Three popular simulation languages are SLAM II, GPSS, and SIMAN. Further information about these and other products are found in references [10] through [14].

VII. SUMMARY

This chapter addressed the need to define the manufacturing system with TMA in mind. It defined the second major element of TMA, manufacturing system control. The system needs to be defined in a way that facilitates controllability.

The manufacturing area is laid out to allow maximum efficiency. It is important that manufacturing subsystems be arranged together in a design that reduces distances, inconvenience, and inflexibility. Methods available to achieve this goal include From-To charting, REL charting, and space-relationship diagramming. Computer programs are available that expedite such analyses and allow for comparative assessment of several candidate layout designs.

Material handling is always an issue. We want to move materials and work-in-process through the facility cost-effectively. The best material handling method is a function of how far and how fast we want, or need, to move material. How these questions are answered guides the investment decision and the sophistication of the handling equipment.

System efficiency is also enhanced through the use of automation. This technology enables repetitive tasks to be performed faster and more consistently. Automation is generally classified as either hard or soft. Soft automation is intended to be flexible. Many factors (for example, programmability) are involved in selecting the most cost-effective automation approach, if any, for a manufacturing system.

In this age of computer technology, a major system definition advantage exists: simulation. Before we invest lots of money in a manufacturing system design that may or may not be what is desired, the candidate design(s) can be evaluated by uniquely developed or commercially available computer simulation programs. This allows the optimal system design to be selected for implementation.

VIII. QUESTIONS

1. Think of several pairs of departments in your organization that deserve an "A" closeness rating. Are they located as close together as they should be? Also, consider pairs of departments that deserve E, I, O, U, and X ratings.

2. Select a list of five to ten departments in your organization that work together frequently. Develop a From-To chart to describe the material handling burden between them.
3. List several examples of material handling equipment used in your organization. Rank them in order of initial investment, from most expensive to least expensive. Now rank them according to unit operating cost, from maximum to minimum. What does this tell you?
4. Recall the material handling principles discussed in this chapter. Think of an example of how each is used in your organization. Think of an example of how each is violated.
5. List several examples each of hard and soft automation in your factory. If there are none, what tasks can you identify that would be suitable for each type of automation?
6. Discuss the differences between DNC and CNC. What are their relative advantages and disadvantages? Which would be most appropriate for each task in your organization?
7. Identify several tasks in your plant that could be handled by robots. What characteristics of these tasks make them especially suited for robotics?

IX. REFERENCES

1. Francis, R.L. and White, J.A.: *Facility Layout and Location: An Analytical Approach*. Prentice-Hall, Englewood Cliffs, New Jersey, 1974.
2. Muther, R.: *Practical Plant Layout*. McGraw-Hill, New York, 1955.
3. Muther, R.: *Systematic Layout Planning*. Industrial Education Institute, Boston, Mass., 1961.
4. Sepponen, R.: *CORELAP 8 User's Manual*. Department of Industrial Engineering, Northeastern University, Boston, Mass., 1969.
5. Seehof, J.M. and Evans, W.O.: Automated Layout Design Program. *The Journal of Industrial Engineering*, Vol. 18, No. 12, 1967.
6. Buffa, E.S., Armour, G.C., and Vollmann, T.E.: Allocating Facilities with CRAFT. *Harvard Business Review*, Vol. 42, No. 2, 1964.
7. Apple, J.M.: *Plant Layout and Material Handling*. John Wiley and Sons, New York, 1977.
8. Muther, R. and Haganas, K.: *Systematic Handling Analysis*. Management and Industrial Research Publications, Kansas City, Missouri, 1969.
9. Azadivar, F.: *Design and Engineering of Production Systems*. Engineering Press, Inc., San Jose, Calif., 1984.
10. Pritsker, A., Alan, B.: *Introduction to Simulation and SLAM II*, 3rd ed. Halsted Press, John Wiley and Sons, New York, 1986.
11. Pritsker, A.A.B., Sigal, C.E., and Hammesfahr, R.D.J. *SLAM II: Network models for Decision Support*. Prentice-Hall, Englewood Cliffs, New Jersey, 1989.

12. Schriber, T.J.: *Simulation Using GPSS*. John Wiley and Sons, New York, 1974.
13. Pegden, D.C.: *Introduction to SIMAN*. Systems Modeling Corporation, State College, Penn., 1987.
14. Hoover, S.V., and Perry, R.F.: *Simulation, A Problem-Solving Approach*. Addison-Wesley, Reading, Mass., 1989.

6

Product Degradation Control

All products have designed-in levels of reliability, safety, and quality. Once the product design is transferred to manufacturing, these inherent levels are degraded through material and workmanship flaws. Consequently, controls must be integrated into the overall manufacturing process to maintain the product reliability, safety, and quality levels as close as possible to the designed-in levels.

This chapter consists of three sections addressing reliability control, safety control, and quality control. The first section addressed the control of product reliability degradation, including prediction, growth, and an analysis technique to aid in identifying the optimal manufacturing system. The second section discusses several analysis methods for identifying critical product safety areas (both product and occupational). The last section discusses techniques (including statistical process control) for controlling product quality during manufacture.

I. INTRODUCTION

This chapter addresses the control of product performance degradation during manufacture. Specifically, the focus is on controlling reliability, safety, and quality degradation. These three performance parameters are critical to a product's success in the marketplace, as discussed in Chapter 3. Admittedly, purchase cost often strongly influences the customer's initial assessment of a product. However, the bottom line is that a product must be able to stand the test of time, and high reliability, safety, and quality are the keys to making this happen.

But remember that we want to focus on controlling the *degradation* of these performance attributes. To do this, we must assume that a product is necessarily designed to have the inherent levels of reliability, safety, and quality that make it desirable. This in itself is generally a poor assumption. All too often, product reliability, safety, and quality are secondary issues in designing and developing new products.

Also, remember that in many cases the basic design itself is flawed. How many times have all of us looked at some simple item and said "Gee whiz, this is a stupid design!" (You can replace "Gee whiz" at your own discretion.) Or the product may be too difficult, or impossible, to manufacture because of its complexity or assembly requirements that surpass the state-of-the-art manufacturing technology. It is hard to imagine a product from either one of these scenarios coming out of a manufacturing system without some type of problem or defect.

When we talk about control, two key points must be kept in mind. First, product degradation during manufacture occurs as a result of defects introduced during the process. Second, no matter how hard one tries to remove and elininate defects, a probability always exists that products with latent and/or patent defects are shipped to customers. Don't forget Mr. Murphy's Law.

Therefore, to eliminate or minimize degradation, we must have all our ducks in a row. This includes, first, ensuring that products are designed with optimal performance safety margins. Second, materials must be specified with proper characteristic tolerances. Finally, manufacturing systems (or subsystems or equipment) must be designed to be reliable and maintainable; to be capable of putting out consistent products; and to be able to be integrated with special inspections, screens, and tests, as appropriate, to detect and remove defective products. It is the manufacturing system equipment that enables a process to be performed.

It is easy to see that a product must be designed to exhibit the performance levels of reliability, safety, and quality required by the marketplace. In the same light, it must be understood that the manufacturing equipment will not necessarily maintain the required parameter levels. The playing field doesn't remain the same. The raw materials change, the operational environment changes, and the equipment parameters change.

For example, as we will see in Chapter 7, equipment maintenance plays a critical role in manufacturing. In an ideal world there is no equipment wear, deterioration, or aging. Unfortunately, in the real world system equipment requires a maintenance strategy and plan.

This chapter focuses on defining and monitoring the system indenture level at which product manufacture actually occurs, in order to minimize degradation that *will* occur. However, once the overall manufacturing process that the system (and its equipment) will perform is defined, it is necessary to determine the optimal system operating conditions. How fast should the machine run? (That is, what should its throughput be?) What temperature should be used for the hot bath? How often should maintenance be done? Where should the crimp be on the product? (How high should that be and how low should that other thing be in the system equipment, or in the product itself?) These are all questions that must be answered, and all can be answered by formal design of experiments (DOE).

By developing and implementing the optimal product and manufacturing system and process design, numerous TMA benefits are derived. Leading the list is increased long-term profitability. Others include

- Design
  -Improved manufacturability
  -Reduced sensitivity to variability
  -Value analysis/engineering implemented
  -Maximized functional performance
- Testing
  -Factor significance (critical vs. unimportant)
  -Engineering priorities
- Capability
  -Optimal procedures
  -Meeting production goals
  -Optimal operational parameters

Figure 1 provides an overview of the DOE process. Depicted are three major steps: (1) planning the experiment, (2) implementing and collecting the data, and (3) performing analysis and interpreting the experimental results. Each step itself has a number of elements.

In planning, perhaps the most important concern is the experimental objective. A major benefit of formal DOE is its introduction of structure and discipline and its requirement to define exactly the question to be answered. After defining an objective, it is necessary to get the proper people involved who can provide valuable input based on their industry experience. It is also important to define all the statistical ground rules to be used in the study, such as the decision rule, test statistic, significance level, and so forth.

In collecting data, good management practice must be used in running the experiment. The results are almost certain to be biased or invalid if the data is not run and collected as defined. The experimental run should reflect a randomized order to minimize data variability due to introduced biases.

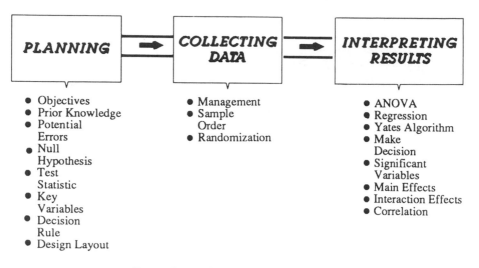

Figure 1   Design of Experiment Steps

In interpreting the results, it is critical to properly identify significant effects that impact the experimental outcome. This includes using proper statistical techniques to determine the statistical significance of results and combining this information with engineering intuition and experience. If practical, replication of results is desirable.

DOE often provides unexpected answers to questions that somebody thought they already knew the answers to. This saves valuable time lost to "shooting in the dark."

We have touched only briefly on the whole area of DOE. There are many texts that provide excellent insight to DOE (for example, [1], [2], and [3]). We strongly encourage the reader to explore the literature available and find a text that is easy to understand and that is at the appropriate level of mathematical detail.

This chapter consists of three main sections addressing product reliability control, safety control, and quality control. We emphasize again that to control these product performance parameters, they must be present in the design before beginning manufacture. The attainment of TMA ensures that these parameters *remain* present in the product design upon shipment to the customer.

II. RELIABILITY CONTROL

A product's designed-in level of reliability is easily degraded during manufacture. High inherent reliability does not counteract the effects of the typically dynamic nature of the manufacturing environment. Both incoming materials and manufacturing system and process parameters change. Such changes may be unnoticeable or may be catastrophically apparent. The thing to remember is that all changes affect product performance in some manner, whether good, bad, or indifferent.

A common method of detecting product changes is by tests or inspections. However, if only a single product passes an inspection or test, it does not imply that all items in the production lot have the same reliability. Inspections and tests are primarily effective in removing patent defects. The factors that degrade product reliability are more often latent defects.

Examples of patent and latent defects are

1. Patent Defects
   a. Broken or damaged in handling
   b. Wrong or no part installed
   c. Faulty part installation
   d. Electrical overstress or electrostatic discharge damage (ESD)
   e. Missing parts
2. Latent Defects
   a. Partial damage through electrical overstress or ESD
   b. Partial physical damage during handling
   c. Material or process-induced hidden flaws
   d. Damage from soldering operations (excessive heat)

Latent defects are identified through the application of various operational stresses. These defects are not readily obvious and they may not exist in all products. This means that they are likely to move into the marketplace undetected and result in early failure when the product encounters an uncharacteristically low overstress situation. The trick is to transform latent defects into patent defects through some type of environmental stress screen. These defects then can be identified and removed by special inspection and/or test.

This problem highlights a significant difference between the philosophical objectives of inspections and tests and the objectives of stress screens. Inspections and tests are intended to pass products, whereas stress screens are intended to fail products.

In dealing with latent defects, it is necessary to implement a manufacturing product reliability control program that complements the design/development reliability program. From such a properly planned degradation control effort, it is possible to maintain product reliability during manufacture.

A. Reliability Fundamentals

But what exactly is product reliability? This is a good question. It is important to understand the meaning of reliability and the benefits to be gained from its control during manufacture.

Reliability is most easily explained by the famous *reliability bathtub*, or life characteristic curve. Figure 2 illustrates this curve [4].

The figure shows the three failure components that comprise the overall product life characteristic curve. These are (1) quality failures, (2) overstress failures, and (3) wearout failures. The distributions of these failure components come together to form the infant mortality, useful-life, and wearout life periods.

The bathtub shape is formed by the decreasing, leveling off, and then increasing hazard rate, or instantaneous failure rate, associated with a product over its lifetime. The high hazard rate during infant mortality is due to latent and patent defects induced during manufacture. The useful-life period reflects a constant hazard rate, or failure rate, that is due to random overstress occurrences, faulty maintenance practices, and/or remaining latent defects. In the wearout period, an increasing hazard rate is evident that results from the gradual physical or chemical change of an item over time causing a decrease in strength. (Keep in mind that the bathtub curve applies to products as big as an airplane or flexible manufacturing system, or as small as a shoe-string.)

Note in the figure that product reliability is inversely related to the hazard rate. This means that as the hazard rate decreases, reliability increases. Although reliability itself is a probability (of success), it is often expressed in terms of a mean time to failure (MTTF) or, more commonly, a mean time between failure (MTBF) (if the product is repairable). The MTBF is derived from the following:

$$\text{MTBF} = \int_0^\infty R(t)\, dt$$

where $R(t)$ is the reliability function.

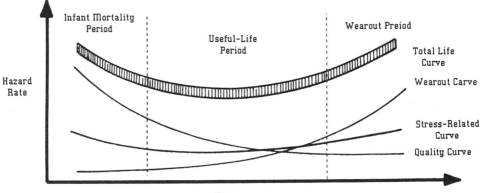

Figure 2   Product Reliability Life Periods

During the useful-life period, the relatively constant hazard rate enables reliability to be described by the exponential distribution. From this, the inverse relationship between the constant hazard rate, or failure rate, and MTBF is derived.

$$\text{MTBF} = \int_0^\infty R(t)\, dt = \int_0^\infty \exp(-\lambda t)\, dt = 1/\eta$$

Note that this relationship applied only during the useful-life period.

Obviously, it is desirable to be in the useful-life period. This is achieved by growing reliability during product development and implementing effective inspections, tests, and screens (ITS) during manufacturing to minimize the infant mortality period. In addition, wearout effects are minimized by a comprehensive maintenance program that addresses short-life parts and components. Proper preventative maintenance actions pull the product back into the useful-life period and delay wearout (discussed in Chapter 7).

Accordingly, the focus is on making the useful-life period as long as possible. From a design perspective, this involves increasing the separation between the probability distributions of stress and strength by derating items (that is, the intentional reduction of stress to strength). From a manufacturing perspective, this involves reducing the scatter of product strength through tighter processes and assurance controls. Finally, from the product user's perspective, this involves controlling application stress variations.

1. Reliability Prediction

A key activity during design and development is to predict the product's inherent reliability. This provides a way to assess any discrepancy between the specified reliability requirement (and lower allocations) and the designed-in level. This activity also provides a way to judge design improvements and alternatives with respect to reliability. In addition, reliability prediction data is useful input into other activities such as reliability growth and qualification tests, design review, logistics and support cost estimates, and general product improvement programs.

Reliability predictions are made at various time during development. A prediction's accuracy depends on the detail and quality of the design information available. As a design progresses from early to detailed stages, more rigorous prediction methods and models are used to reflect the greater level of definition.

Remember that reliability failures are stress-related failures. These occur during a products' useful-life period and are based on three key assumptions: (1) they occur randomly, (2) they occur independent of each other, and (3) they occur at a constant mean rate.

Also, keep in mind that all failures are mechanical in nature. This applies to both mechanical parts (such as gears) *and* electronic parts. Electronic parts are more commonly associated with reliability methods because a standardization of part types exist. This enables large data-collection efforts, leading to the development of generic reliability prediction algorithms.

Accepting the fact that all failures are actually mechanical (since electrons do not fail), it is a good idea to have some understanding of the strength characteristics of an item or material. There are two general theories for viewing the strength probability distribution for materials [5]. The first states that the strength of a material is determined by its weakest point. The distribution of strength is then determined by the lowest value of samples of points in the material. This is represented by the extreme value distribution.

The second theory states that the weaker points in a material receive support from surrounding stronger points (an averaging effect occurs). The distribution of strength is the mean value of all points. This is represented by the normal distribution.

Both theories state that materials can and often do exhibit strengths well below their theoretical capacity. This happens because of the existence of imperfections, or defects, induced by manufacturing. The probable strength is related to the effect of the imperfection creating the greatest reduction in strength. A reliability failure results when a random overstress occurs that exceeds the probable strength. In other words, when stress exceeds strength, failure occurs. Using reliability prediction, one can obtain an estimate of how often (on the average) an overstress failure may occur (that is, MTBF).

Table 1 identifies the hierarchy of reliability prediction techniques used to assess reliability based on the strength-stress interactions. The techniques accommodate different analysis requirements and the availability of detailed data as the product design progresses.

Remember that reliability predictions are best used for comparison purposes. Use and look at reliability prediction data for what it really is: a figure-of-merit of performance. However, the more accurately a prediction model presents the actual operational application, the more credible is the figure-of-merit.

Table 1 Hierarchy of Prediction Techniques

Similar Equipment—Do broad comparison of similar items of known reliability.

Similar Complexity—Compare similar type items based on design complexity.

Similar Function—Evaluate based on previously demonstrated correlations between operational function and reliability.

Part Count—Evaluate based on the number of parts, in each of several part classes, included in the item.

Stress/Strength—Evaluate based on the individual part failure rates, and considering part type, operational stress level, and derating characteristics of each part.

---

2. Growing Product Reliability Prior to Manufacture

An important consideration during product development is reliability growth. As stated earlier, one of the first tasks performed during product development is reliability prediction. The actual product reliability level, however, will be lower than that predicted.

A reliability development and growth test (RDGT) program enables a design's reliability to grow to the level required. Growth occurs during the test, which consists of operationally stressing the product to force out problems. A program results from analyzing failures to determine root causes, developing and implementing sound corrective action, and verifying that the failure has been effectively eliminated.

Formal RDGT is performed for a full-scale preproduction article that reflects the final product configuration. It is also manufactured (as closely as possible) using the process(es) associated with the planned manufacturing system. The intent is to get the product's reliability level at or near that predicted, which is done by eliminating or minimizing failure sources through a rigorous test program. Supporting this RDGT effort must be a comprehensive failure, recording, analysis, and corrective action (FRACA) system (as discussed in Chapter 9).

Knowing that reliability *will* degrade during manufacturing, it makes sense to apply RDGT to ensure that the product design itself meets reliability requirements. Attention can then be focused completely on controlling reliability degradation occurring as a result of the manufacturing system and raw material variations.

The RDGT planning and implementation activity aims to meet a specified level of product reliability at the minimum cost. This requires that the RDGT program reflect a proper balance among desired hardware reliability, test cost, and test time [6].

As with all projects, a detailed management plan is essential. The RDGT plan defines the cumulative test time required to grow to the specified or targeted MTBF, the number of test units, and the anticipated test time per unit. The plan also defines details of the growth model, reliability starting point, growth rate, and test schedule.

A popular model for planning the reliability growth process is the Duane model [7]. The Duane model assumes that if a uniform RDGT program is implemented, reliability growth maintains a linear relationship with time. Also, the rate of the growth is dependent on the rigor of the test, analyze, and fix (TAAF) program implemented. Figure 3 illustrates this model.

Two key variables depicted in the figure are the reliability starting point and the growth rate. The reliability starting point, $MTBF_0$, is typically between 10 and 20 percent of the predicted MTBF. A conservative value should be used to ensure adequate funding and time to complete the RDGT.

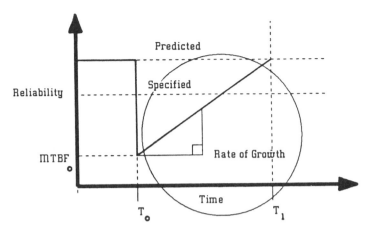

Figure 3   Duane Model Illustration

The value of the growth rate typically varies between 0.1 and 0.6. A growth rate of 0.1 is expected in programs in which no specific consideration is given to reliability. Typically, a conservative growth rate of 0.3 to 0.4 is planned in all cases. Although rates of 0.5 and 0.6 are theoretically possible, they are seldom achieved.

Note that the time required for an RDGT can be lengthy (a year or more). This does not mean that RDGT should be avoided, but that one should explore the many RDGT program trade-offs possible to reduce test time while not reducing test integrity. Common trade-offs to reduce RDGT schedules involve increasing FRACA program rigor and the number of dedicated test units.

B. Manufacturing System Modeling

As an integral part of designing and evaluating a manufacturing system, it is desirable to estimate the product reliability degradation that is going to occur. This is done by performing product reliability manufacturing degradation analysis (PRMDA) (see Figure 4).

The analysis is composed of two fundamental activities. The first consists of modeling the manufacturing system with a detailed flow diagram depicting each step of the overall manufacturing process that is performed by the system. This includes the ITS integrated into the overall process, as appropriate.

The second activity involves exercising the model to estimate the number of defects introduced into the overall process, removed from the process, and outgoing from the process. These estimates are then used to derive a degradation adjustment factor that is multiplied with the predicted product MTBF to provide an estimate of system outgoing product MTBF.

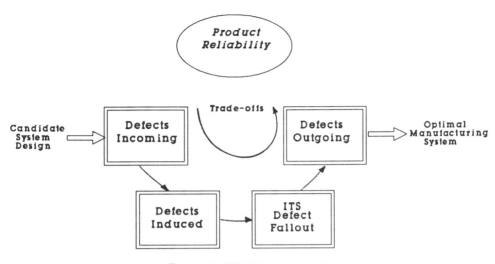

Figure 4   PRMDA Approach

The analysis enables knowledgeable trade-offs to be made between the type and quantity of ITS to be integrated into the manufacturing system and process. This makes it possible to optimize the overall effectiveness of the manufacturing system design and minimize the product defectivity rate transferred to the consumer markets.

At this point, we should distinguish between reliability control and what is done as part of quality control (discussed later in this chapter). With reliability control, we are able to minimize the number of failure-causing (particularly, near-term) product defects that make it to the marketplace. For the most part, these defects are not readily observable; they are latent. Thus, the need for special ITS.

With quality control, we are able to minimize the number of readily observable product defects that make it to the marketplace. The key word here is *minimize*. Statistical process control, for example, is designed to minimize the number of defects that can be seen by the customer. Certainly, zero defects is the ultimate goal. Reliability degradation control makes this goal more practical.

Integrating ITS into the manufacturing system requires knowledge of the quantity and type of defects expected in the product (at various levels). It is the reliability, degrading defects (that is, latent defects) that we are concerned with in this section.

The quantity and type of latent defects that are introduced into a product depend on several factors [8]. These factors include

- Design Complexity—The quantity and type of parts and interconnections used in the product. Increased complexity creates more opportunities for defects.
- Part Quality Level/Grade—The quality levels of parts or materials.
- Operational Environment—The stress conditions to which the product will be exposed in the actual use environment.
- Manufacturing System/Process Maturity—New systems and processed require learning time to identify and correct planning and process problems, train personnel, and establish vendor and process controls.
- Packaging Density—Product assemblies with high part and interconnect density are more susceptible to system/process, workmanship, and environmentally induced defects.
- Manufacturing System/Process Control—Good control and monitoring reduce the number of defects that are introduced into the product.
- Workmanship Standards—Stringent and enforced workmanship quality standards minimize workmanship defects introduced into the product.

Even though we design what is viewed optimistically as a product-reliability-conscious manufacturing system, keep in mind that ITS is not perfect. At each step of manufacture where ITS is applied, defects will escape to the next process step and new opportunities for introducing defects will be created.

1. Defect Removal Planning

As with other engineering activities, defect removal planning for the manufacturing system must begin early in the design phase. Successful defect removal depends strongly on knowing the product, the manufacturing system, and the overall process. The following factors should be considered:

1. The type of potential defects in the product
2. Experience data available for similar products (for example, in composition, construction, and maturity)
3. Information sources, including
   a. Item historical performance
   b. Supplier/vendor performance and certification
   c. Qualification test data
   d. Supplier-provided test data
   e. Incoming inspection data
   f. ITS records for previous manufacturing programs
   g. Reliability growth test results
4. The need for cost-effective integration of ITS into the manufacturing process
5. Project management activities
   a. Establishing objectives/goals

b. Preparing a written plan
c. Obtaining planning estimates of defect density
d. Selecting and placing of ITS

2. PRMDA Technique

The PRMDA technique allows various manufacturing system configuration alternatives to be evaluated. This means that the system configuration providing the best relative benefit-cost is identified before any significant resource investment. The following formula can be used to determine this benefit-cost figure-of-merit.

$$RBC = (DAF)(ECD - EIC)$$

where RBD is the relative benefit-cost, DAF is the reliability degradation adjustment factor, ECS is the expected cost savings (rework or repair costs times number of defects removed), and EIC is the expected implementation cost (hardware costs plus inspection or test costs). The DAF is derived (later in this section) based on the removal of expected latent defects.

Several key factors in assessing the benefit-cost include

- Outgoing defectivity rate
- Incoming defectivity rate
- Types of ITS relative to product characteristics
- Screen effectiveness
- Test effectiveness
- Inspection effectiveness
- Product environmental design limits
- ITS facility costs
- ITS monitoring costs
- Material scrap quantities
- Product rework costs
- Product field repair costs
- Failure analysis costs
- Product volume

Many of the above factors are fixed for a given corporation, particularly those associated with basic costs. The PRMDA adjusts the multipliers for these fixed costs by exercising the model in light of various trade-off scenarios (that is, manufacturing system ITS configurations).

Earlier in this chapter, PRMDA was defined as consisting of two fundamental activities. The first activity focuses on modeling the manufacturing system and the processes supported by its. The second activity focuses on exercising the model to estimate product reliability degradation. These activities are addressed in the following paragraphs.

3. PRMDA Activity One

In developing a model, a systems engineering approach is used. The manufacturing system is viewed as essentially a black box. There are defects coming into the box, induced in the box, falling out of the box, and going out from the box to the customer. Figure 5 illustrates the highest level of this modeling approach.

The level at which a system is modeled is a judgment call. However, modeling must be to a level that depicts individual steps of the overall manufacturing process(es) and ITS locations integrated within it. Figure 6 depicts some lower levels of manufacturing system modeling. Note the identification of the PRMDA parameters. These are the number of latent defects in materials entering the system (D_I), the effectiveness of each ITS $(E_1, E_2 . . . E_n)$, the number of latent defects in materials entering each process (or step) (D_{IN}), the number of latent defects going out for each process (D_O), the latent defect fallout from each ITS (D_F), and the number of latent defects going out of the system (D_R).

4. PRMDA Activity Two

To exercise the model, it is necessary to estimate the values of the various PRMDA parameters. The first parameter to estimate is D_I. This is done via manipulation of the Chance Defective Exponential Formulation [8].

$$\lambda_{s(t)} = \lambda_o + (D_I/N)(\lambda_{Id})[\exp(-t^*\lambda_{Id})]$$

where: $\lambda_{s(t)}$ is the actual item failure rate
         λ_o is the predicted item failure rate
         D_I is the number of incoming latent defects
         N is the quantity of material or number of parts

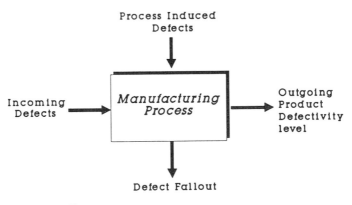

Figure 5   Systems Modeling Approach

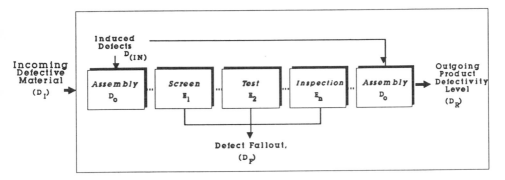

Figure 6   Detailed Systems Model

λ_{ld} is the assumed latent defect failure rate
t is the amount of exposure (e.g., time, length, area, and volume)

The parameter D_1 is derived for this formulation by the application of several key (and conservative) assumptions. First, the average failure rate of a defect in the marketplace is greater than one failure per thousand t in order to be considered an early failure, or latent defect. From this, η_{ld} is estimated as 0.001 defect per t. Second, for products undergoing RDGT, the actual product failure rate is at least 50 percent of that predicted. For products not undergoing RDGT, the product failure is rate is at most 30 percent of that predicted. Applying these assumptions allows D_1 to be derived as

$$D_1 = (1000)(\lambda_o)(N) \quad \text{(with RDGT)}$$
$$D_1 = (2300)(\lambda_o)(N) \quad \text{(without RDGT)}$$

The second parameter to estimate is E for each inspection or test in place to detect the latent turned patent defect. Inspection effectiveness is a function of two possible implementation scenarios: manual and automated. Automated inspection methods are typically the most cost-effective, and values for E of 0.95 and higher are appropriate. For manual methods values for E of 0.7 to 0.9 are generally appropriate.

A more exact formula for determining E value is derived from the following [9]:

$$E = (Eb)(Et)(Ec)(Ee)$$

where: E is the inspection effectiveness level
Eb is the baseline error probability under ideal conditions
Et is a factor to account for level of training or experience
Ec is a factor to account for inspection complexity
Ee is a factor to account for the inspection environment

Test effectiveness is defined as the ratio of patent defects detectable by a defined test procedure to the total possible number of patent defects present. Although stress screens are effective in transforming a latent defect into a detectable failure, removal of the failed condition depends on the capability of the test procedures used to detect and localize the failure.

It is important to ensure that tests have detection efficiencies as high as is technically and economically feasible. Difficulty in simulating functional interfaces accurately or an inability to establish meaningful acceptance criteria reduces test effectiveness. Table 2 provides example values of test effectiveness.

Screen effectiveness is defined as the ability of the screen to precipitate a latent defect to a detectable state, given that a defect susceptible to the screen stress is present. A basic premise of stress screening is that with specific stresses applied over time, latent defect failure rates of materials are accelerated over those that would occur under normal field operating stress conditions. For example, by subjecting an electronic product assembly level to accelerated stresses (such as rapid temperature cycling and random vibration), latent defects are precipitated to early failure.

More severe stresses accelerate both failure mechanisms and the rate of defect failure. However, care must be taken not to implement destructive screens, but to stress the subject item only within its design limits.

The most popular screens are vibration and temperature cycling. In general, vibration screens are more effective for precipitating workmanship-induced defects, and thermal screens are more effective for part defects. There are also classes of defects that respond to both vibration and thermal screens.

It is advantageous to exercise the product operationally during the screen. Also, performing the screen at the lowest possible product level (that is, part, intermediate, or product) reduces the cost of rework or repair. However, finding problems at higher levels of assembly generally costs less.

An important issue to remember is that the applicable product level must be tested before entering a screen. Otherwise, one cannot determine whether the defects were precipitated by the screen or were present in the product level (as patent defects) before the screen.

Table 2 Test Effectiveness, E

Test Type	Detection Effectiveness
Manual Go/No-Go	0.85
Operational	0.9
Automated Functional	0.95

The screen effectiveness value is derived from the following formula:

E = (SS)(DE)

where: E is screen effectiveness
         SS is the screening strength (see Table 3)
         DE is the detection effectiveness (this value is derived from either
         the inspection or test effectiveness criteria, discussed earlier, de-
         pending on which is applicable)

Once the above PRMDA parameter values are defined, it is possible to estimate the number of defects going out to the marketplace. This involves suming the D_1 values moving into each process and the outgoing D_o value going out to each defined ITS as applicable. The D_F is determined by multiplying its E value to derive the outgoing D_1 into the next process, as applicable. These calculations continue through the model until the D_R value is estimated.

It is now possible to estimate the product reliability degradation occurring during manufacture [10]. The following formula provides this estimation:

$MTBF_a$ = $(MTBF_i)(DAF)$

where: $MTBF_a$ is the achieved reliability (resulting from the degradation
         forces)
         $MTBF_i$ is the predicted reliability
         DAF   is the degradation adjustment factor

The degradation adjustment factor is derived from

DAF $- D_1/(D_1 + D_R)$

This value can be used in the relative benefit-cost equation defined previously.

5. PRMDA Example

Consider the manufacture of a hand-held electronic product that costs $1,000. Ten thousand units are planned to be manufactured. Each unit consists of several integrated circuits, semiconductors, and resistors. The predicted MTBF for the item is 26,315 hours. A temperature cycling screen with a range of 80 °C and a 10 °C per minute rate of change and power cycling is to be performed at the board level (SS = 0.9). Test effectiveness is 0.95. The cost of a repair is $200 (considering inspection time, repair time, materials, field service costs, and so on), and the cost of implementing the screen into the manufacturing system and process is $10,500 (considering thermal cycling chamber, test equipment, accessories, inspection time, handling time, and so on). Figure 7 illustrates the model.

The following are derived:

DAF   = 825/(825 + 120) = 0.87
$MTBF_a$ = (26,315 hours)(0.87) = 22,894 hours

Table 3   SS Values (Adapted from DOD-HDBK-344)

Temperature

Number of Cycles	Temp Rate of Change, °C/Min.	$\dot{T}$	80	100	120	140	160	180
6								
	$\dot{T}>20$:	5	.70	.75	.79	.82	.84	.86
		10	.90	.93	.95	.96	.97	.98
		15	.97	.98	—	.99	—	
		.99						
8								
	$\dot{T}>15$:	5	.80	.84	.87	.90	.91	.93
		10	.96	.97	.98	—	.99	—
		.99						
10								
	$\dot{T}>15$:	5	.87	.90	.92	.94	.95	.96
		10	.98	—	.99	—		
		.99						

Random Vibration

Duration Per Axis (minutes)	Acceleration Level (grams)													
	0.5	1.0	1.5	2.0	2.5	3.0	3.5	4.0	4.5	5.0	5.5	6.0	6.5	7.0
5	0.007	0.023	0.045	0.07	0.10	0.14	0.18	0.22	0.26	0.30	0.35	0.39	0.43	0.47
10	0.014	0.045	0.088	0.14	0.20	0.26	0.32	0.39	0.45	0.51	0.57	0.63	0.68	0.72
15	0.021	0.067	0.13	0.20	0.28	0.36	0.44	0.52	0.60	0.66	0.72	0.77	0.82	0.85
20	0.028	0.088	0.17	0.26	0.36	0.45	0.54	0.63	0.70	0.76	0.81	0.86	0.90	0.92

Incoming Material

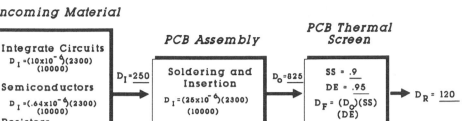

Figure 7   Example System Model

ECS   = ($200)(705) = $141,000
EIC   = $10,500
RBC   − (0.87)($141,000 − $10,500) = $113,535

III. SAFETY CONTROL

Controlling product safety during manufacturing is a critical issue. Clearly, product designs should never reach the manufacturing floor if they exhibit a significant risk to the customer. The legal liabilities can prove very costly, both in dollars and in reputation. This section also addresses the manufacturing system as a product, since it also should be designed to minimize product safety degradation and present no impending danger to its operations.

Safety is an important function of TMA. Safety engineering is a broad discipline, but it focuses primarily on three fundamental areas: (1) products, (2) systems, and (3) occupational environments. Within these areas, extended focus is on the customer and on the personnel working in the manufacturing environment.

Product safety is important to the customer; system and occupational safety are important to the workers in the manufacturing facility. All three elements are a function of design and must receive priority attention from management. The rationale is simply the fact that corporate liability exists for each of these areas. Furthermore, there is absolutely no justification for knowingly endangering persons using your product or making your product. The key is to know who you are designing for and then use this information to develop a design that can be used safely and effectively.

To ensure that safety is adequately controlled, it is imperative to apply engineering design evaluation techniques that address both product and manufacturing system safety. Is is also important to develop a corporate safety program plan to improve worker morale and productivity relative to their perspective of manufacturing system and occupational safety. (This ties into worker motivation, discussed in Chapter 4.)

A. Safety Assessment

One can generally expect a large number of safety issues to be present during the development of new products or manufacturing processes and systems. Discovering and addressing these issues require that a formal risk assessment be performed.

A risk assessment, or hazard analysis, provides insight into the risks or potential hazards involved with the item in question. From this assessment, responsible management decisions can be made concerning the perceived risk(s). For example: Is the perceived risk reasonable or acceptable? Can the risk be made acceptable through the use of protective measures, or is it completely unacceptable (which would dictate redesign or discontinuing the development effort)?

Obtaining the information required to answer these questions involves several activities. First, it must be determined whether or not the probable risks are acceptable. Second, the justifiable level of funding for accident prevention measures must be determined. Third, comparisons must be made between risks and hazard rates for similar existing products or processes.

Many techniques exist to identify safety hazards and rank them in terms of their criticality. Fault tree analysis (FTA) and failure mode, effects, and criticality analysis (FMECA) are two popular and powerful techniques. These analyses provide a rationale for addressing those hazards that, when eliminated or reduced, provide the greatest impact in reducing risk. In addition, these analysis techniques provide:

- A method of selecting optimal designs
- A method for assessing and documenting failure modes and their effects
- Early visibility of hardware and human-machine interface problems
- A ranking of potential failures relative to severity of effects and probability of occurrence
- Identification of single failure points critical to operation or performance
- Criteria for test planning
- Quantitative and qualitative data and information for use in reliability, maintainability, and logistic analyses
- A basis for design and location of performance monitoring and built-in test capabilities
- A basis for evaluating design, operational, or procedural changes

In terms of overall cost and project life cycle, it makes sense to eliminate or reduce design hazards as early as possible during the design/development cycle. This requires beginning the risk assessment during the design concept phase and then updating the analysis with current information as the development effort progresses.

A risk assessment/hazard analysis identifies the existence of potential hazards and their probability of occurrence. The adequacy of proposed hazard controls are also determined. To determine whether or not the hazard safeguards provided are adequate, a verification effort is required. The verification effort provides assurance that

1. The subject item and corresponding operating procedures mitigate the potential for a safety incident.
2. The subject item exhibits no hazardous characteristics not forseen by analysis.
3. All potentially hazardous characteristics are controlled.
4. The requirements of industry codes and standards are satisfied.

Verification may be performed during design and development or randomly during production or operation.

Typically, design safety verification is achieved by one or more of the following: technical evaluation, inspection, demonstration, or test. Technical evaluation involves taking an in-depth look at the theoretical basis for design, as well as the reference documentation (for example, code(s) or standard(s) that served to guide the basic design formulation). Technical evaluation is often substituted for other means of verification that may be too difficult, time-consuming, or costly.

Inspection provides verification through visual (or other sensory), possibly enhanced, detection of workmanship or material characteristics that affect the existence of a hazardous condition. Hazards may also be detected through some form of measurement or simple manipulation. Demonstration is a trial conducted to show that a specific operation can be accomplished safely, that a product operates safely, or that a material contains or lacks a certain property. Tests are more detailed than demonstrations in that specific measurable parameters must be met. A test is often used to verify that values for a certain operational parameter fall (or do not fall) within specified limits and that various operational parameters do not cause a failure, damage, or hazardous condition.

To be completely safe, a candidate product or manufacturing system or process must have no potential for causing injury or damage under any circumstances. Unfortunately, no such product or process exists. Therefore, some risk must always be expected and accepted during the normal use of any design. How much risk is acceptable depends on the benefits derived from usage. Therefore, when risk is greater than that deemed acceptable, the subject design must be considered unsafe. In this case, the design must be made safe through some compensating or correction action before being introduced for use.

1. Fault Tree Analysis

Fault tree analysis is a graphical technique applied to products and systems for modeling the various faults that lead to the occurrence of a defined, undesired hazardous event. The analysis allows a product or system to be analyzed in the context of its environment and operation to find all possible ways in which the undesired event can occur.

The fault tree itself provides a visual representation (see Figure 8) of the logical interrelationship between a specific failure event, or basic fault, and the ultimate effect is has upon the subject item. A basic fault reflects a specific part or component hardware failure(s), human error(s), or any other pertinent event(s) that can lead to a higher undesired event. Note that this analysis technique is different from FMECA (to be discussed later in this section) in that is addressed human error. FMECA is limited to evaluating hardware and functional failures.

A key part of FTA is the definition of the top undesired event to focus on. The top event gradually reflects a complete or catastrophic failure of the item under consideration. Careful choice of the top event is important to the success of the analysis. If it is too broad, the analysis cannot adequately address a specific problem. If it is too specific, the analysis cannot adequately address all significant issues contributing to a specific problem.

The FTA process is an interactive approach to identifying basic faults and establishing their criticality. The power of the fault tree comes from its qualitative and quantitative ability to assess criticality. A fault tree structure reflects a series of logic gates that serve to permit or inhibit the passage of fault logic up the tree to the top event. Figure 9 presents the variety of logic symbols used in a fault tree diagram.

Thus, through qualitative and/or quantitative evaluation, the fault tree provides an assessment of the probability of reaching the top undesired event. For example, visually examining the fault tree diagram for the types of logic gates that predominate gives an indication of the relative ease with which the top event can occur.

This analysis power enables corrective action recommendations (both design and operational) to be readily formulated and prioritized to enhance product or system safety cost-effectively. The analysis technique is most beneficial when it is implemented early in design conceptualization and is

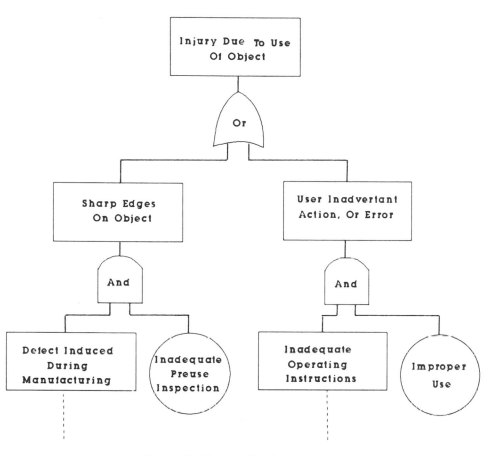

Figure 8   Example Fault Tree Diagram

updated throughout the design and development phases in light of design changes.

Fault tree analyses involve four major steps [11]:

Step 1: *Construct the fault tree diagram.* This involves defining the top event of the tree using terminology that will encompass the lesser events, individually or collectively. Next, beginning with the top event, paths leading to each succeeding lower level of detail are developed. This requires having a thorough enough understanding of the product or system to visualize all the events that could conceivably take place as a result of malfunctions, failures, or human error. The diagram expands as a progression of events through the logic gates.

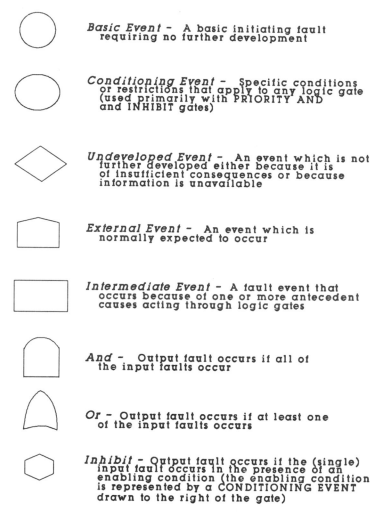

Basic Event - A basic initiating fault requiring no further development

Conditioning Event - Specific conditions or restrictions that apply to any logic gate (used primarily with PRIORITY AND and INHIBIT gates)

Undeveloped Event - An event which is not further developed either because it is of insufficient consequences or because information is unavailable

External Event - An event which is normally expected to occur

Intermediate Event - A fault event that occurs because of one or more antecedent causes acting through logic gates

And - Output fault occurs if all of the input faults occur

Or - Output fault occurs if at least one of the input faults occurs

Inhibit - Output fault occurs if the (single) input fault occurs in the presence of an enabling condition (the enabling condition is represented by a CONDITIONING EVENT drawn to the right of the gate)

Figure 9   Fault Tree Logic Symbols

Step 2: *Collect basic fault data.* This involves compiling statistical data based on experience (either specific or industry generic) to develop a failure probability for each basic unit.

Step 3: *Evaluate the diagram.* This involves determining which combinations of failures are most likely to cause the top event of the tree (that is, which failure paths are dominant). In doing this, the criticality of each failure is determined by calculating its failure impact on the top event. A secondary evaluation goal is to determine the probability of the top event occurring;

however, the use of this value should be limited to serving as a benchmark to assess design and operational improvements.

Step 4: *Formulate corrective action recommendations*. This involves prioritizing the basic faults relative to their criticality and defining corrective action to reduce the most critical faults' impact on the occurrence of the top event of the fault tree diagram.

2. Failure Mode, Effects, and Criticality Analysis

Failure mode, effects, and criticality analysis is a technique for systematically evaluating and documenting the potential impact of each functional or hardware failure on operational success, personnel and system safety, system performance, maintainability, and maintenance requirements. Not that human factors are not considered in this analysis technique.

As with FTA, the result is a ranking of each potential failure mode relative to the severity of its effects. This enables appropriate corrective actions to be taken to eliminate or control high-risk items.

The analysis is inductive. This means that the analyst induces failure to each item at the lowest applicable level of product, system, or process hierarchy. FMECA actually serves as a paper test prior to expensive hardware assembly and configuration in a test fixture. From a knowledge of the predominant failure modes of each element, the analyst traces up through the various levels of the subject item to determine the effects a failure will have on the subsequent indenture levels and, ultimately, output performance.

This differs from FTA, which is a deductive technique. As stated previously, FTA uses a top-down approach that assumes that the top hazardous event has occurred. From this, all the failure modes and basic faults contributing to the occurrence of the top event are defined.

The FMECA is begun early in the conceptual phase to support the evaluation of candidate designs and to provide a basis for establishing corrective action priorities. The analysis plays a key role in design reviews from concept through final hardware development and, ultimately, technology transfer. In addition to the obvious benefits that FMECA provides to actual hardware design, FMECA aids in defining special test considerations, quality inspection points, preventative maintenance actions, operational constraints, useful-life factors, and other pertinent information and activities necessary to minimize failure risk. Note that all recommended actions resulting from FMECA should be evaluated and formally dispositioned by appropriate implementation *or* by documented rationale for no action.

The analysis consists of two fundamental parts. First, the failure modes and effects of each indenture level component must be identified. Second, the criticality of each failure mode must be calculated.

The FMECA documents all probable failures in a product, system, or process within specified ground rules. These include the analysis approach, the lowest level of indenture to be analyzed, sources of data and information describing the subject item and failure of components within, and include general statements of what constitutes a failure. Every effort should be made to identify and record all ground rules and assumptions before beginning the analysis. These ground rules and assumptions must be expanded thereafter as necessary.

The following steps are performed during the analysis [12]:

Step 1: *Define the item to be analyzed*. This involves gathering comprehensive information including identification of internal and external interface functions, expected performance at various indenture levels, item restraints, and failure definitions.

Step 2: *Construct functional and reliability block diagrams*. This involves developing diagrams that illustrate the operation, interrelationships, and interdependencies of functional components of the design.

Step 3: *Identify and evaluate failure modes*. This involves assessing all potentially significant failure modes of each component at the lowest appropriate indenture level. Next, the effects on the next higher indenture levels are determined. Finally, the worst-case and effect is defined relative to the following severity classifications:
- Category 1—Catastrophic, a failure that may cause loss of life
- Category 2—Critical, a failure that may cause severe injury, property damage, or loss of function
- Category 3—Marginal, a failure that may cause minor injury, property damage, or degraded function performance
- Category 4—Minor, a failure of low severity resulting in unscheduled maintenance or repair

Step 4: *Identify failure mode detection methods and compensation provisions*. This involves establishing the means for isolating failures through the use of inspection criteria or design features such as built-in test equipment (BITE).

Step 5: *Determine design corrective action priorities*. This involves calculating the criticality of each failure mode and ranking them.

Step 6: *Formulate specific correction actions*. This involves identifying specific corrective actions (either design or procedural) for the most critical failure modes.

The FMECA is documented on a worksheet such as that shown in Figure 10. The worksheet is straightforward and facilitates conduct of the analysis.

The top blocks on the worksheet provide background information. The remainder of the columns are delineated below.

References Number—The unique part identification number.
Identification—The part name.

FAILURE MODE, EFFECTS, AND CRITICALITY ANALYSIS

| ITEM | IDENTIFICATION NO. | | SHEET ____ OF ____ |

ITEM DESCRIPTION/MISSION — DATE

INDENTURE LEVEL — REFERENCE DRAWING — PREPARED BY — REVISION NO.

REFERENCE NO.	IDENT.	FAILURE MODE (FM)	FAILURE EFFECTS (FE)			COMP. PROV.S	SEVERITY CLASS	λ_p	FE PROB. (β)	FM RATIO (α)	OPER. TIME (t)	FM CRIT. (C_m)	PART CRIT. (C_R)	REMARKS
			LOCAL	NEXT HIGHER LEVEL	END									

Figure 10 FMECA Worksheet

Failure Mode—The part potential failure mode under consideration.

Failure Effects—The failure effects of the failure mode at subsequently higher indenture levels.

Compensation Provisions—The design features or procedural routines in place to offset the effects of the failure mode.

Severity Class—The worst-case failure mode classification.

λp—The part failure rate.

Failure Effect Probability, β—The conditional probability that the failure effect results in the identified criticality classification. If unknown, use 1.0.

Failure Mode Ratio, α—The fraction of the component's total failure rate due to the failure mode under consideration.

Exposure, t—The exposure of the subject item (e.g., hours).

Failure Mode Criticality, Cm—The calculated criticality of the failure mode under consideration:

$$Cm = (\beta)(\alpha)(\lambda p)(t)$$

Part Criticality, Cp—The cumulative sum of the failure mode criticalities calculated for the part.

B. Safety and the Worker

In addition to product and manufacturing system safety, occupational safety plays a key role in the company, particularly in regard to worker morale. It is desirable to have a safe and healthy work environment. Establishing a safe work environment allows the production floor personnel to concentrate on their work productivity and quality. This in turn leads to more efficient and effective production and to high-quality, cost-competitive products.

Obviously, the manufacturing environment does not have to be absolutely safe for products to roll out the door. An unsafe work environment, however, may directly result in reduced product throughput and lower quality. In addition, a lack of safety may increase product cost to the consumer due to direct and indirect accident costs.

A worse scenario develops when workers perceive that management is not concerned about their safety. The result is that the workers may not care if the product goes out the door "right." Management can relieve worker preoccupation with avoiding injury by implementing a well-defined occupational safety program. The plan for such a program identifies operational critical safety issues, serves to establish good worker morales, and ultimately establishes a quality product base and a healthy management operation.

1. Human Factors

Human factors include providing work environments that foster effective procedures, work patterns, and personnel safety and health, and minimizing factors that degrade human performance or reliability. Thus, the intent is to design manufacturing systems and equipment in such a way that operator workload, accuracy, time constraints, mental processing, and communication requirements do not exceed operator capabilities. In addition, designs should minimize personnel and training requirements within the limits of time, cost, and performance trade-offs.

In short, it is necessary to optimize the man-machine interface (MNI). Many examples exist where the safety or operational effectiveness of complex systems have been compromised by defects in the design of the MMI and have rendered the systems difficult or impossible to operate.

This leads us to the realm of ergonomics. Simply defined, ergonomics is the science of designing based on the characteristics of the intended users of the design in its final form and in its intended environment. This means applying the "principle of user-centered design" [13].

User-centered design promotes four fundamentals:

1. Directly observe human beings and their behavior, supported by systematic investigations of human experience.
2. Apply both empirical analysis and evaluation.
3. Modify the product to fit the user.
4. Achieve the best possible match for the greatest number of people as far as is reasonably practical within constraints of cost and time.

The end result is a design that reflects an allocation of functions to personnel, equipment, and MMI to achieve the required sensitivity, precision, time, and safety; the required reliability of system performance; the minimum number and level of skills of personnel required to operate and maintain the system; and the required performance in a cost-effective manner.

The following is a list of generic MMI issues to consider, as applicable, in developing a user-oriented design [14].

*   Satisfactory atmospheric conditions, including composition, pressure, temperature, and humidity
*   Range of acoustic noise, vibration, acceleration, shock, blast, and impact forces
*   Protection from thermal, toxicological, radiological, mechanical, electrical, electromagnetic, pyrotechnic, visual, and other hazards
*   Adequate space for personnel, their equipment, and the activities they are required to perform, under both normal and emergency conditions
*   Adequate physical, visual, auditory, and other communication links between personnel, and their equipment, under both normal and emergency conditions
*   Efficient arrangement of operation and maintenance areas, equipment, controls, and displays

- Provisions for ensuring safe, efficient task performance under reduced and elevated gravitational forces with safeguards against injury, equipment damage, and disorientation
- Adequate natural or artificial illumination for the performance of operations, control, training, and maintenance
- Safe and adequate passageways, hatches, ladders, stairways, platforms, inclines, and other provisions for ingress, egress, and passage under normal, adverse, and emergency conditions
- Provision of acceptable personnel accommodations, including body support and restraint, seating, rest, and so forth
- Provision of nonrestrictive personal life support and protective equipment
- Provision for minimizing psychophysiological stress effects of fatigue
- Design features to ensure rapidity, safety, ease, and economy of operation and maintenance in normal, adverse, and emergency environments
- Satisfactory remote handling provisions and tools
- Adequate emergency systems for contingency management, escape, survival, and rescue.
- Compatibility of the design, location, and layout of controls, displays, work areas, maintenance accesses, and storage provisions with the clothing and personal equipment to be worn by personnel operating or maintaining systems or equipment

The most effective way to avoid MMI problems is to use standardized MMI design criteria. Such standardization should encompass controls, displays, marking, coding, labeling, and arrangement schemes. Remember to base internal design standards on recognized national or international sources.

2. Safety Program Planning

Using knowledge gained from safety management theory and practice, it is possible to define the basic elements of a successful safety program. This includes addressing the key elemental motivators that have the strongest effect on employee morale and attitude.

Keep in mind that in defining a tailored program plan, it is always easier to remove task elements than it is to add elements at a later date. Therefore, include items in the initial plan even if you strongly suspect that they will be removed from the final program plan.

Basic elements of a safety program include the following elements.

Management Commitment—Program commitment is displayed in two ways. First, a written management statement of safety policy is prepared and signed by top management (that is, the president or chief executive officer and displayed to all corporate employees). Second, the commitment is backed by actions. This requires that top management take part in safety-related activities to display interest and ensure that the policy is adhered to.

Assignment of Responsibility—A single person is assigned responsibility for overall management and direction of the program. This includes having the authority to make decisions and direct actions. Typically, this person is a safety director in a large organization, but it can be any effective manager with adequate training.

Supervisor Responsibility—Supervisors are held responsible and accountable for the safety activities and accident experience within their respective departments and they are given the authority to meet this responsibility. As necessary, the supervisor must seek assistance to solve safety problems from either in-house or out-of-house sources.

Supervisor Training—Supervisors are educated to create a safe environment. They learn how to deal with people, recognize hazards and safety concerns, perform accident investigations and identify corrective actions, train employees, and know their importance to the overall success of the safety effort.

Hazard Identification—A project for hazard identification is implemented. There are several techniques to accomplish this and the simplest is probbly periodic inspection of a manufacturing operation in accordance with a customized checklist. When a hazard is identified, a corrective action is formulated and followed up. Employee input is another good way to communicate hazards. State-of-the-art technology using video recorders provides another way to observe hazardous operations.

Safety Committees—A committee made up of management and labor representation is developed and given specific responsibilities and objectives. Top management involvement enhances the effectiveness of such a group significantly. The objective of the committee can be general in scope, such as ongoing review of loss history, hazardous input, hazard control input, and so on. It can also be strictly on a per-project basis, such as to research and develop new procedures, operational controls, and so forth. The committee focuses on product issues, liability, or even vehicular functions of the operation.

Employee Training—Safety training is provided to employees and periodically reinforced. Initial new employee orientations include at least the basic safety concerns of not just one job, but of the whole corporation. Specific safety training for individual jobs is presented by a competent trainer. Ongoing training of employee groups is needed to reinforce specific concerns and may be related by loss history or degrees of operational exposures. A 5-minute safety talk given once a week is a feasible alternative. This is quick, effective, and takes a minimal amount of time. It has a consistent effect on reinforcing management concern.

Accident Investigation—Establishing a specific procedure to investigate and document each accident and incident (close calls with no loss) is critical. Corrective action is then formulated to prevent accidents from recurring.

Accident Analysis—Data analysis involves studying past loss history to identify trends of accidents by body part, cause, equipment, location, department time of day, shift, and so forth. This aids in focusing on those areas
that are most significantly impacted by corrective action.

First Aid/Medical Facilities—Having the ability and the means to take care
of injured persons is basic to any safety program. As the minimum, a first
aid kit is required, along with persons who are trained in first aid. Specific
plans for addressing a major medical emergency are necessary.

IV. QUALITY CONTROL

The control of product quality naturally encompasses many important elements. First of all, the product must be designed for quality. Next, the process
by which it is manufactured must be designed and controlled for maximum
product quality. Finally, the entire process must be managed for adherence
to these quality goals.

When we speak of quality control, however, we are usually referring to
the statistical monitoring and control of process quality and adherence to
standards. This activity is more accurately named statistical process control
and will be the focus of this section.

We will assume that a product has been adequately designed for the purpose it is to serve. Its tolerances have been adequately specified, and a manufacturing process has been selected that can produce the item correctly and
economically. Our job now will be to make sure that the process continues
to function properly and, if it deviates, to get some understanding of the
nature of the problem.

A. On-Line vs. Off-Line

The statistical analysis of a process can be viewed from two different vantage
points: on-line and off-line. The on-line perspective assumes that we are
currently manufacturing the product and examines each item or batch as it
rolls off the assembly line. We can make conclusions about the process parameters and implement feedback techniques to correct problems as they
develop, hopefully before they become significant. Naturally, this on-line
perspective lends itself ideally to the production process. It is usually referred to as process monitoring.

Statistical process control can also be implemented off-line. This involves
examining large batches of product, possibly long after they have been produced. The techniques in this case are time-independent, because it is generally not known which items were produced first. Conclusions to be drawn
concern the entire batch and whether or not it conforms to standards. Little

can be inferred as to why a particular batch may be inferior. This off-line implementation may be used to monitor a manufacturing process if on-line techniques are not feasible, but it more naturally lends itself to decision making between independent units in the production process. For example, an inspection department may periodically test a day's worth of product to check up on a production department. More frequently, however, this off-line control strategy is employed when acquiring items from a vendor or supplier. Entire shipments may therefore be accepted or rejected on the basis of their overall quality, and the vendor is left to worry about the cause of any degradation. Since these techniques are used to decide whether a batch is to be accepted or not, it is known as acceptance sampling.

B. Statistics and Probability

The fundamental science of statistical process control is statistics. Associated with statistics is the concept of probability. In statistical process control, we use statistics to calculate probabilities, and we use the probabilities to derive our conclusions.

Probability is a measure of our knowledge about some aspect of the world. When our knowledge is complete, the probability of any outcome or event is either one or zero, depending on whether the event has occurred or not. When our knowledge is incomplete, the probability is somewhere between zero and one. The classic example is tossing a coin. It will land with either heads or tails up, and while it is in the air, spinning in an incalculable manner, we just do not know how it will land. We assign a probability of one-half to each of the possible outcomes, because we have no way of knowing which will occur, or even if one is more likely to occur than the other.

Of course, if we had good enough sensors, detailed enough math models, accurate data on air currents in the room and the precise torques imparted to the coin as it was flipped, along with a fast enough computer, we could have predicted exactly how the coin would land. The probability of the correct outcome would be adjusted to one, the other to zero. Without this knowledge, however, we must be satisfied with a 50 percent chance of each outcome.

Once the coin lands, the question has been resolved. We now have more knowledge, because we can look at the coin on the table and see which side is up. After the event has occurred, all probabilities are either one or zero. Again, the degree of knowledge determines the numerical probability that we assign to an event.

This is how statistics are used in process control. Just like the tumbling coin, a manufacturing process has some exact state, and some precise number of nonconforming units will be produced. Before the fact, however, we do not know this number and will have to assign probabilities. Statistics collected on past events are used to predict the probabilities of future events.

If X represents an event that may or may not occur, we will use the designation Pr(X) to denote the probability of X occurring. If we define an experiment as a process whose outcome is not yet known, then the set of all possible outcomes is called the event space of that experiment. Since one of the outcomes definitely must occur, the sum of the probabilities of all events in the event space must be unity.

A probability distribution is a function or rule that assigns probabilities to the events in an event space. If we are to deal with an unknown process mathematically, we must know what the distribution is. How is it arrived at?

Basically, three methods exist for determining the probability distribution of an event space. One is based on reasoning. Each possible outcome is analyzed in terms of its causes, and the probability of its occurring is calculated based on the probabilities of its precursors. This approach is generally not feasible, since any process complicated enough to require a statistical treatment will proceed according to largely unknown mechanisms.

A more generally useful method of determining a probability distribution is to collect data on past experiments. If 1,000 experiments are run, and the number of occurrences of each possible outcome are tabulated, a distribution can be calculated. A simple histogram of results is a form of distribution, although not particularly useful. If an equation can be fit to the data, of the form

Pr(X) = some function f(X)

we have a useful tool for further mathematical analysis. The only problem with this technique is that it requires that the process under consideration not change with time. That is, the probabilities of each outcome must remain constant into the future, or the function that was fit to the data is no longer valid.

The third method is to force the event space into some preselected distribution. If we can design an experiment that is guaranteed to have some known distribution, we are home free. This approach has intuitive appeal and is a favorite method in process control.

How can this be done? Fortunately, mathematics supplies us with the precise tool we need in the form of the central limit theorem. In mathematical terms, this is stated [15]:

For a population having mean μ and finite standard deviation σ_p, let $\overline{X}$ represent the mean of n independent random observations. The sampling distribution of $\overline{X}$ tends toward a normal distribution with mean μ and standard deviation

$$\sigma = \frac{\sigma_p}{\sqrt{n}}$$

In plain words, this tells us that whatever the distribution is of the raw data coming out of our process, we can force the data into a normal, or Gaussian, distribution by looking at samples of the data.

Suppose our process is spitting out a stream of data of unknown distribution. The first datum is X_1, the next is X_2, then X_3, X_4, and so on, up to the last piece of data, X_m, where m is some very large number.

If we take some subset of this data, say the first n values, we can calculate a subset, or sample, average. We call this $\overline{X}_1$. The mean of the next n values is called $\overline{X}_2$, and so on. We will be able to calculate m/n of these $\overline{X}$ values.

The central limit theorem tells us that is n is large enough, the distribution of the $\overline{X}$s will be normal. This is very useful! The only question left to be answered is, how large is large enough? The precise answer to this question depends on the precise distribution of the original data and how close we wish to come to a normal distribution. The larger the n, the closer we come to true normality. In most engineering applications, 4 is considered a lower limit on sample size. However, if data is plentiful and measurements are cheap to make, 10 is a much better size to use. The sample size in any actual implementation will be based on the cost and availability of the data and the precision required in the analysis.

1. The Normal Distribution

The famous normal distribution, also known as the bell curve or Gaussian distribution, is easy to handle. In probability terms, it is expressed as

$$\Pr(X) = \frac{1}{\sigma\sqrt{2\pi}} \exp \left[-\frac{1}{2} \left(\frac{x-\mu}{\sigma} \right)^2 \right]$$

where μ is the process mean and σ is the standard deviation.

To calculate the probability of any outcome X, we need to know only the process mean, the standard deviation, and the above formula. Remember, in our process control situations, we will be using sample means as our data, and the X in the equation must also be an average of a sample. Finding μ and σ is a simple matter of collecting data and calculating them, and then assuming that the process remains stable in the future. The formulas for μ and σ are

$$\mu = \overline{X} = \frac{1}{n} \sum_{i=1}^{n} x_i$$

$$\sigma = \sqrt{\frac{\sum\limits_{i=1}^{n}(X_i - \mu)^2}{n-1}}$$

where n is the number of samples.

It should be noted that these formulas would give the exact values of the process mean and standard deviation only if infinite data were collected. However, a reasonably large set of data is the best we can do and is generally good enough.

C. Process Capability Analysis

Before we begin monitoring a process to see if it remains within our required level of quality, we must determine if it begins within the required level of quality. This is referred to as capability analysis.

Any process in the real world is subject to random variables. These are minimized to the greatest extent possible, but some will always remain. Raw materials will have some inherent variability, as will the machines that process the material. People operating the machines will perform their duties in slightly varying ways, both from person to person and from hour to hour for one person. Some amount of variability, of a purely random nature, must be accepted.

If we have designed our process intelligently, these variations will be kept to a bare minimum and will follow a random, unpredictable distribution. The cumulative effect of all the random variations within a process will have an even less predictable effect on the outcome of the process. However, through the use of sample means, we can force this effect into a normal distribution, after which it can be measured and quantified.

According to the mathematics of the normal distribution, any outcome is possible. The farther the outcome is from the mean, however, the less likely it is to occur. Of course, there are practical limits involved. For instance, if we are tracking the weights of items, we will not get a negative value. But in general, vast deviations from the nominal value are always possible and will occur, however infrequently.

Our job in capability analysis is to determine just how infrequently these large deviations do occur. We can define process capability as that range of variation in a product characteristic that will include almost all of the product produced. For instance, if our goal is to fill a soda can with 12 ounces of soda, what range of values will include the amount of soda in almost all the cans we fill? The smaller this range of variations is, the better.

But we have left ourselves open again: What do we mean by "almost all" of the product? If we define "almost all" as 50 percent, we will be able to claim a fairly small, and therefore fairly good, process capability. In our example, we might say that 50 percent of all soda cans have between 12.01 and 11.99 ounces of soda. If we chose 90 percent as our definition of "almost all," we might have to admit to a looser capability, since we would have to include more of the mistakes and deviations. In our example, the best 90 percent of the cans may vary between 12.5 and 11.5 ounces. To include 100 percent of the cans, we would need to extend our limits to the practical limits of the situation; in this case, from zero to the volume of the can itself.

In practice, it is common to define "almost all" of a process as anything within plus or minus three standard deviations of the mean. In a normal distribution, the sum of the probabilities of all outcomes between the plus and minus three sigma limits comes out to 99.73 percent. If we choose the $\pm 3\sigma$ limits and claim that the resultant limits are our process capability, we are saying that 0.27 percent of our product will be outside those limits. Other definitions are of course possible, and some companies are now using plus and minus six sigma as their capability limits, which includes 99.9999998 percent of the product. Table 4 lists percentages for other sigma values.

Process capability limits are often confused with tolerances. Although they look alike, both having an upper and lower limit, they are completely independent concepts. Tolerances are what we want or need, while capability limits are what we have. If the capability limits and the tolerances are equal, we are in a satisfactory situation; 99.73 percent of our product will be within tolerance, which is quite good. If our capability limits are outside of our tolerances, that is not so good. It means that our process is not precise enough to produce product that is mostly within the required limits.

The best possible situation is when the capability limits are well within the tolerances. This means that more than our 99.73 percent is satisfactory, and even if the process were to degrade, we would still be in good shape. Of course,

Table 4 Percentages for Sigma Values

Process Limits	Percent of Distribution Included
$\pm 1\sigma$	68.26
$\pm 2\sigma$	95.46
$\pm 3\sigma$	99.73
$\pm 4\sigma$	99.9937
$\pm 5\sigma$	99.999943
$\pm 6\sigma$	99.9999998

if the capability range were much smaller than the tolerance range, we might be wasting time and money on excess quality that is going to waste. A common rule of thumb is that capability range should be around 70 to 80 percent of the tolerance range.

Sometimes it will happen that the process cannot be tightened to within the tolerance range. In this case, there are several options, none of them good. One is to use 100 percent inspection on the output, separating the good from the bad. Another is to redesign the process from scratch, purchasing more precise equipment or hiring workers with greater skill. A last resort is to change the tolerances by redesigning the product. This requires extensive changes in many departments and will not make the process engineer very popular in the overall organization, but it sometimes is the only approach.

A commonly used measure of the relationship between process capability and desired tolerances is known as the process capability index, or C_p. This is defined as the difference between the upper specification limits (USL) and the lower specification limit (LSL), divided by 6 times the process standard deviation (the range of process values, which includes $\pm 3\sigma$ sigma of the process); which is to say, 99.73 percent of the process output.

$$C_p = \frac{USL - LSL}{6\sigma}$$

If exactly 99.73 percent of the process output falls within the tolerances, then the C_p will be exactly 1.0. As discussed above, this represents a good situation, but leaves no margin for error. Similarly, if 99.73 percent of the process lies within a range smaller than the tolerance range, the C_p will be greater than 1.0, and we will be in a better situation. If the C_p is less than 1.0, it means that to include 99.73 percent of the process output, we need a range greater than the specified tolerance range, and we are in trouble. To follow the rule of thumb given above, an acceptable C_p would be in the range of 1.20 to 1.4.

D. Process Monitoring

Process monitoring is the on-line version of statistical process control, in which we collect data and make inferences about a process and its quality while the process is occurring. The goals are to detect process degradation as quickly as possible and correct any problems before an excessive amount of substandard product is produced.

One of the possible degradations that we will be monitoring is an increase in the process capability, as defined above. That is, we will be watching to see if the range of variation that includes some standard percentage of our product increases. An example of this would be if our soda cans, formerly filled to between 11.5 and 12. 5 ounces, now contain anywhere between 11.0 and 13.0 ounces of soda. If something of this nature occurs, we will want to know as soon as possible.

Another of the possible degradations that we will be watching for is a shift in the mean. This can occur without a loss in process capability. That is, the size of the range that includes most of the product is the same, but the location of the range has changed. For example, our soda cans no longer contain between 11.5 and 12.5 ounces, but now contain between 9.5 and 10.5 ounces.

The basic technique of process monitoring is the control chart. This is a graphical tool that tracks product quality over time and gives visual evidence of problems, as well as intuitive clues as to their possible origin. Many forms of the control chart exist, and we will look at only a few of them.

A control chart is a graphical "hypothesis test." The assertion, known in mathematics as the *null hypothesis*, is that the process capability has not changed. Each time a new piece of data is added to the chart, this assertion is tested. The result of each test will be either to accept the null hypothesis, or to reject it.

In its simplest form, a control chart consists of three horizontal lines and a vertical scale. These lines are called the center line (CL), the upper control limit (UCL), and the lower control limit (LCL). The vertical scale reflects possible values of the quality characteristic we wish to track. In the example from the previous section, we were considering ounces of soda in a soda can. In that case, the scale would go from zero to the volume of the can, so that any possible value of our variable could be plotted (see Figure 11).

The horizontal axis of the chart can be thought of as time or, more precisely, measurement number. Measurements will be taken at regular intervals, so that each new entry in the chart will be one unit to the right of the previous entry.

Remember, the measurements that we plot are not individual values, but mean values of samples of some size n, preferably at least four. As long as all points on the chart stay within the UCl and LCL, we consider the process to be "in control." That is, nothing is going wrong, the process capability is remaining stable, and the mean is not drifting. However, if points begin to wander outside of the control limits, the process is becoming "out of control," meaning that something has gone wrong.

How do we calculate the locations of the limits on our chart? Basically, two phases are involved, known as a base period and a monitoring period. The sequence of steps in control chart development are outlined below.

1. Base Period
    a. Select a quality characteristic to measure.
       Select a confidence level (in our case, $\pm 3\sigma$). Select a sample size, n

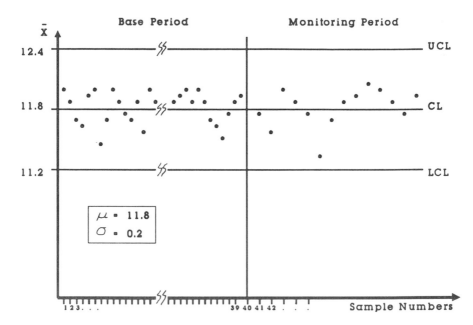

Figure 11 Typical Control Chart for a Variable

(at least 4). Select *N*, the number of samples in the base period (20 is considered a good minimum, 50 is better).

b. Take $N * n$ measurements, and calculate $N\,\overline{X}$s. Calculate $\overline{X}$ the average of the sample means. This is an estimate of the process mean and will be the center line of the control chart. Calculate σ, the standard deviation of the $\overline{X}$s. Calculate UCL = $\overline{\overline{X}} + 3\sigma$. Calculate LCL = $\overline{\overline{X}} - 3\sigma$.

c. Draw the CL, UCL, and LCL at the appropriate location on the control chart. Plot the $N\,\overline{X}$s on the chart, in the order in which they were measured.

d. Examine the chart, and decide if the process is in control. If it is in control, there will be no points outside control limits and no "trends" in the data. If the process is in control, go on to monitoring period.

2. Monitoring Period

a. At regular intervals, take *n* measurements. Calculate a new $\overline{X}$ from the new measurements. Plot the new $\overline{X}$ on the chart. If it is within the control limit, and no trends are appearing, the process is remaining in control.

b. Repeat.

Notice that we have two different situations that signal an "out of control" process. One is an $\overline{X}$ outside of the control limits, and the other is any sort of trend in the data. Each of these is worth further discussion.

An $\overline{X}$ value that falls outside of the control limits is the easiest red flag to interpret. But what does this actually mean? We constructed our control limits according to our definition of capability. If we assumed $\pm 3\sigma$ limits, then 99.73 percent of our $\overline{X}$s would be inside those limits. That implies a finite change of any one $\overline{X}$ being outside the limits. But that chance is extremely slight. In our base period, with anywhere from 20 to 50 $\overline{X}$s, the chance of one falling outside the limits, due merely to chance in the distribution, is negligible. If an $\overline{X}$ is outside the limits, it is far more likely that there is something wrong with the process, that is was out of control from the start, and that there is no point in progressing any further.

We can be slightly more liberal during the monitoring period. If we have been coasting along with an in-control process for a long, long time, and suddenly one $\overline{X}$ falls outside the limits, what is the probability that nothing is wrong? Since 99.73 percent of an in-control process should fall inside the limits, only 0.27 percent can be expected to fall outside. That amounts to about one value in 370. If we have had hundreds of $\overline{X}$s inside the limits before one falls outside, we might let it slide. But if two points in, say 100 values fall outside, we will definitely want to investigate the process and look for problems. If your control chart is based on other than $\pm 3\sigma$ limits, of course, these numbers would be different.

What about trends in the data? The assumption of a normal distribution calls for more than just control limits. It also requires that the data be randomly distributed between those limits. This means approximately equal points above and below the center line, more near the center line than near the control limits, and no discernible pattern to the data. Typical patterns that might appear, signaling an out-of-control process, would include a slow, steady drift away from the center, a cyclic distribution, or two distinct populations (see Figure 12). Any of these patterns indicate that the process is not as it appears and should be investigated.

What might cause these trends? The control chart alone cannot pinpoint the problem, but it can give certain insights. For example, a sudden jump in the distribution of points would indicate a sudden change in the process, perhaps caused by a new supplier of raw materials, a new operator on a machine, or a minor failure of some component in the process. A gradual change in the distribution, on the other hand, would signal possible wear of a tool or component, or some other gradual degradation of a process element.

A cyclic trend in the data would indicate that some sort of cyclic influence that was not anticipated is acting on the process. Environmental effects often

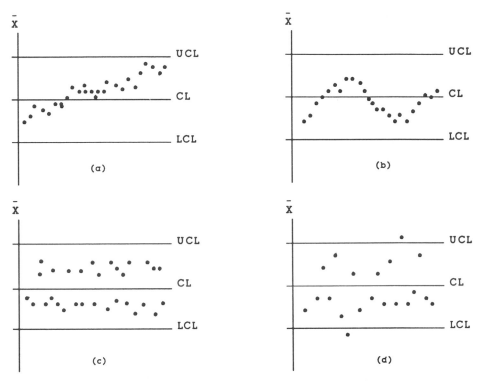

Figure 12   Control Chart Out-of-Control Conditions

fall into this category, such as temperature variations from day to night or from season to season. If the period of the cycle is weekly or daily, it could be based on personnel factors, such as midweek burnout, Monday hangovers, or lunchtime and coffee-break effects. Maintenance schedules may also cause cyclic variations in the process.

   Whatever cause is ultimately determined for the trend, it should be eliminated, preferably by removing the source of the variation. If this is not possible, the data should be segregated into groups, each of which will have its own trend-free chart.

1. Types of Control Charts

The procedure outlined above for developing a control chart is just the bare bones of the process. Many different embellishments exist for specific situations and can be found in any text on quality control. The most important of the specific control charts, and their uses, are discussed here.

2. Control Charts for Variables

A variable can be defined as a quality characteristic with a numeric value, such as weight, length, diameter, resistance, strength, hardness, and so forth. Control charts based on variables are developed when the actual value of the quantity is important. If more than one characteristic is to be tracked, separate control charts must be developed for each one.

In tracking variables with a control chart, we are actually trying to understand a random stream of numbers. The mathematics of so-called stochastic systems can show that a random number stream has many degrees of freedom and can be viewed in many different ways. In practical terms, this means we will generally want to develop more than one control chart for each variable. In normal practice, two charts are used, one to track the central tendency of the variable, and one to track the variability of the variable. For the central tendency, we use an $\overline{X}$ chart, and for the variability, we use either a range chart or a sigma chart.

The $\overline{X}$ chart is the most basic type of control chart and is precisely as described above. The quality characteristic is measured in samples, and each sample's mean, or $\overline{X}$, is plotted against the average and $\pm 3\sigma$ values of the $\overline{X}$s.

The range chart follows the same basic philosophy, but is calculated quite differently. For each sample, we calculate the range R, or the largest value minus the smallest value. If we started with N samples, we will have N ranges. We next calculate the average of these ranges, R, and the standard deviation of the ranges, $R\sigma$. We then construct a control chart with R as the center line and $\pm 3\sigma$ as the upper and lower control limits.

Now we have two blank charts, an $\overline{X}$ chart and an R chart. For each sample, in both the base and the monitoring periods, we plot both the $\overline{X}$ and R of the sample. Superficially, the charts look very much alike. In reality, though, they tell us very different things. In-control and out-of-control is defined the same for both of charts: all points should be between the control limits, with no trends. But an out-of-control indication in the $\overline{X}$ chart tells us that the central tendency of the process is changing, whereas a problem in the R chart tells us that the range is changing. Of course, if the range gets smaller, we could be experiencing an improvement in the process, but this is still something that we would wish to notice.

The third chart is the sigma chart. This performs the same function as the R chart: it tracks the variability of the process. The advantage of the sigma chart, though, is that all data are used. A range chart, using only the largest and smallest values in a sample, ignores what the rest of the values are doing. In small samples, say five or less, this is fine. But in larger sample sizes, we could be ignoring valuable information. The sigma chart is created in exactly the same way as the R chart, except that it uses the average standard deviation of the samples, rather than the average range, as the center line, and $\pm\,3\sigma$ of these standard deviations as control limits. In this way, all data is used, and subtler changes can be detected. The only disadvantage of the sigma chart over the R chart is that it requires many more calculations.

In actual practice, calculations of $\overline{X}$, R, and sigma charts are performed with tables, and actual standard deviations of ranges and sigmas need not be determined. Tabulated values, based on sample size n, give factors that are used to determine center lines and upper and lower control limits.

3. Control Charts for Attributes

An attribute is defined as a quality characteristic that has one of two values: good or bad. This is a useful way of judging conformance to specifications when the actual value of a variable is of no importance, as long as some maximum and minimum have been satisfied. An example would be the length of a screw that is long enough to mount a nut on, but not so long that is causes interference. In other circumstances, the quality characteristic being tracked may have no numerical value at all, such as a resistor that has either been mounted on a circuit board, or has not.

Attributes are further divided into two classes: defects and defectives. A defect is a negative quality characteristic, such as a scratch in a surface, a tear in fabric, or a missing component. The important point in defect tracking is that any one manufactured item may have many defects. A defective, on the other hand, is a single item that has too many defects to be considered usable. For most situations, focus on either defects or defectives may be chosen, depending on the preferences and goals of the quality engineer.

Unlike control charts for variables, attribute-based control charts do not use the normal distribution. Since they are not based on variables with numeric values, the normal distribution no longer applies. Control charts for defectives are based on the binominal distribution, and charts for defects are based on the Poisson distribution.

The most common type of control chart for defectives is called a p-chart. In this case, p stands for proportion. We still use samples of some size n, but instead of calculating an average, we calculate the proportion of defectives in the sample. For example, if a sample of 100 items has 28 defectives, the p value for that sample would be 0.28. The choice of sample size n is based on the proportion of defects expected in the long run. Samples should be large enough so that each sample contains at least on defective; otherwise, the calculations will be too coarse to detect the effects of process variations.

In the monitoring period of a p-chart, N samples of n units are inspected, and a p value for each sample is calculated. The average p for all samples, $\overline{p}$, is the center line of the chart. The control limits, based on the binomial distribution, are calculated from

$$\bar{p} \pm 3 \sqrt{\frac{\bar{p}\,(1 - \bar{p})}{n}}$$

This formula gives limits that include 99.73 percent of the population, as before. The p values for each sample are plotted, as in variable control charts, and the p values for each monitoring period sample are plotted similarly. Interpretation is the same as for variables. The only difference is that there is no need for any kind of range or sigma chart, since we are no longer dealing with numbers, but merely attributes.

Control charts for defects are similar, except that the equations are different. That is because the distribution that governs defects is the Poisson distribution. As before, samples of n units are inspected. Now, however, rather than counting defective units, the defects per unit are counted. If defects occur frequently, a sample may consist of a single unit, with n equal to 1. However, if defects are rare, it is wise to define a sample large enough to ensure several defects per sample.

The standard defect chart is called a c-chart, which tracks the count of defects per sample. N samples are inspected, each with some count c of defects. The average number of defects in all N samples is $\bar{c}$ which becomes the center line of the chart. Based on the Poisson distribution, 99.73 percent of the population will be between the limits calculated as

$$\sqrt{\bar{c}} \pm 3 \sqrt{\bar{c}}$$

As usual, the N cs of the base period are plotted on these limits, and the process is judged to be in or out of control. As long as it remains in control, the monitoring period continues with a new c for each new sample, which is subsequently plotted. Again, there is no need for any range or sigma chart.

Many other control charts exist for defects and defectives, as well as for variables. But these give the general idea of what control charts look like and what they can accomplish.

E. Acceptance Sampling

Acceptance sampling is the off-line version of statistical process control, and it is used to infer quality information about a large quantity of items that have already been produced. The goal is to infer the overall quality level of the batch without having to examine every item in the batch.

Acceptance sampling is traditionally an activity performed by a receiving department before accepting a shipment from a supplier. At this point, it is too late to monitor the process that produced the shipment, so the only question to be answered is: Is the quality of the shipment sufficiently high?

This question could be answered by simply examining every item in the shipment, a process called 100 percent inspection. In some cases such inspection is unavoidable, but generally it is too expensive in terms of time, money, and labor. Furthermore, it is not significantly more accurate than inspecting a small sample of the batch and using statistics to infer the overall quality level. Also, because acceptance sampling involves less handling than does 100 percent inspection, it involves less handling damage, which can be a significant advantage in some situations. For items that can be inspected only by destructive testing methods, acceptance sampling is the only feasible method of quality assurance.

Acceptance sampling like all quality control activities, is based on probability and statistics, specifically on probability distribution. In a sampling situation, the important variables are defined as follows:

N = the number of items in the batch
y = the number of defective items in the batch
n = the number of items that we will inspect
x = the number of inspected items that we will allow to be defective

In this sample plan definition, we will accept or reject the entire batch of N items based on the result of our inspection of the n items in the sample. The n items are selected at random from the batch. Naturally, the greater the ratio of N to n, the more savings we will enjoy in comparison to 100 percent inspection.

We have in mind some value x. If x or fewer of the n items we inspect are defective, we will conclude that the batch is of sufficient quality to be accepted. If more than x of the n items are defective, we will reject the entire batch. This means that we are using the x/n ratio to infer information about the y/N ratio, based on our knowledge of the probability and statistics of the situation.

The statistics of this particular situation are based on the hypergeometric distribution. This is the same distribution that governs the selection of numbers in a lottery or the drawing of names from a hat. Its most interesting feature is that the probability of each successive item drawn from the batch having some characteristic (winning lottery number, defective item, or so forth) is different from the probability of the previous item, since the proportion of items remaining in the batch has changed with the removal of the previous item.

Using the definitions of variables given above, the hypergeometric distribution can be used to calculate the probability of any number x of the n inspected items being defective:

$$\Pr(X) = \frac{\binom{y}{x}\binom{N-y}{n-x}}{\binom{N}{n}}$$

$$\text{where } \binom{a}{b} = \frac{a!}{b!(a-b)!}$$

That is, the number of ways x out of y bad items can be selected, times the number of ways $n - x$ out of $N - y$ good items can be selected, all divided by the number of ways n of N items can be selected for inspection.

Using this definition, what is the probability of accepting the batch described above? We will accept as long as the number of defectives in the sample is no longer than x, so we can say that

$$\text{Pr(acceptance)} = \text{Pr}(0) + \text{Pr}(1) + \cdots + \text{Pr}(x)$$

The sample plan described above is an example of a single sampling plan. Another type of plan is the double sampling plan. In this type of plan, we have two sample sizes, n_1 and n_2. We also have two acceptance numbers, x_1 and x_2, and two rejection numbers, r_1 and r_2. We still have only one batch size, of course, N.

In this double sampling plan, we begin by selecting a sample of n_1 items from the batch, and we inspect them. If x_1 or fewer of the n_1 items are defective, we accept the batch. If r_1 or more of the n_1 items are defective, we reject the batch. However, if the number of defectives in the sample, call it d_1, is greater than x_1 but smaller than r_1, we defer making a decision and take another sample. This sample has size n_2, which is often equal to n_1, but need not be. Now we inspect the second sample and determine the total number of defectives from both batches, d. If d is less than or equal to x_2, we accept the batch; if it is greater than or equal to r_2, we reject the batch. In a double sample plan, r_2 must equal $x_2 + 1$, or there would be another indecision zone. Often r_2 is set equal to r_1, but it could be any value greater than or equal to r_1.

This same line of reasoning can be used to develop a multiple sample plan, in which we use any number m of sampling levels. We will have m sample sizes, $n_1, n_2, \ldots, n_m$. Again, it is standard to make these sizes equal, but that is not necessary. We will also have $x_1, x_2, \ldots, x_m$ and $r_1, r_2, \ldots, r_m$, where r_m must equal $x_m + 1$ to ensure that a decision will be made in the mth sample, if there has not been one made before.

Why do we have these complex plans? Single, double, and multiple sample plans all have certain advantages and disadvantages. The single plan is obviously the easiest to implement, since it has only one step. On the other hand, adding levels of sampling decreases the amount of overall sampling. That is, a double sample plan of the same discriminating power as a single sample plan will, in the long run, require less items to be inspected. A multiple sample plan of equal power will require even less inspection.

Exactly how much inspection will be required by a particular sample plan? For a single sample plan, the inspection level is fixed in advance, at n units per batch. However, in the double and multiple sampling plans, the total sampling burden will depend upon the quality of the batch. This burden should be considered when choosing a plan, to determine if the extra complications of the multilevel sampling plan are justified by the savings in sampling cost.

It easily can be seen that the sampling burden is a function of the quality of the batch. Consider a double sampling plan, defined by some n_1, n_2, x_1, x_2, r_1, and r_2. If the batch to be inspected is absolutely perfect, the number of defective items in the first sample will always be zero. Since x_1 will be greater than zero, the batch will always be accepted after the first sample.

However, if there are a very small number of defectives in the batch, we will occasionally need to take a second sample, and the average samples per batch will increase. The worse the batch gets, the more often x_1 will be exceeded and the higher the average will go. Eventually, if we continue to assume worse and worse batch quality, we will begin to reject batches. At medium quality, it will take both samples to reject. But at very poor quality, with a high proportion of defective units in the batch, we will find ourselves rejecting frequently on the first sample. In the extreme case, if the batch is completely defective, we will always reject on the first sample, all n_1 will be bad, and n_1 must be greater than r_1.

This average, in the long run, of the number of items that must be inspected is called the average sample number, or ASN. For a single sample plan, ASN must equal n, whereas for a double sample plan, ASN varies with average batch quality, starting at n_1 for very high-quality batches, peaking between n_1 and $n_1 + n_2$ for medium-quality batches, and dropping again to n_1 for extremely poor-quality batches. Calculating ASN requires us to predict the probability of accepting a batch of given quality, which we shall investigate next.

1. Probability of Acceptance

Suppose we have a batch of items of some quality level q, where q is defined as the fraction of items in the batch that are in some way defective. Suppose further that we have some sampling plan. Will we accept or reject the lot? This depends partly on chance, of course, since we have no way of knowing, before the fact, which of the items we will inspect. The n items to be inspected are chosen at random from the batch. The best we can say is that for a given quality q and a given sampling plan, there is some probability, P_a, of accepting the batch.

The fact that there is a probability involved implies that there is some element of risk. We may reject a batch that is really quite good, merely because we happen to select all the bad items for inspection. Or, even worse, we may accept a horrible batch, merely on the chance selection of the few good items. These risks are the price we pay for reducing our inspection cost through acceptance sampling, rather than using 100 percent inspection.

This price is generally well worthwhile, especially if we take care to minimize the risks involved. For this reason, we need to be able to calculate our probability of accepting a batch of any given quality.

To calculate this P_a, we can use the hypergeometric distribution. An alternative, however, is to use a different formulation known as the Poisson distribution. This distribution is somewhat simpler, and tables exist that can eliminate the need for calculation. The disadvantage of the Poisson distribution in calculating probabilities of acceptance is that it ignores the change in the proportion of nonconforming units in the batch as items are removed. This is a minor problem, though, as long as n is much smaller than N.

To illustrate the calculation of P_a, an example is presented below.

2. Example

A shipment of 100 delicate instruments has just been received. We are using a single sampling plan which specifies that we inspect 10 units and accept the shipment if no more than 3 of those inspected are defective.

Although we do not know it, there are 5 defective instruments in the shipment. What is the probability that we will accept the shipment? If we continue to use this plan on all shipments, what will our average sampling burden be?

To calculate the exact solution, we must use the hypergeometric distribution described above. For this example, the variables are

$N = 100$
$y = 5$
$n = 10$
$c = 3$

where c is the number of defectives we will allow in the sample. We must calculate Pr(no defectives) + Pr(one defective) + Pr(two defectives) + Pr(three defectives).

$$\Pr(x) = \frac{\binom{y}{x}\binom{N-y}{n-x}}{\binom{N}{n}}$$

$$\Pr(0) = \frac{\binom{5}{0}\binom{95}{10}}{\binom{100}{10}} = 0.58375$$

$$\Pr(1) = \frac{\binom{5}{1}\binom{95}{9}}{\binom{100}{10}} = 0.33939$$

$$\Pr(2) = \frac{\binom{5}{2}\binom{95}{8}}{\binom{100}{10}} = 0.070219$$

$$\Pr(3) = \frac{\binom{5}{3}\binom{95}{7}}{\binom{100}{10}} = 0.0063835$$

$$\text{so that } P_a = \sum_{x=0}^{3} \Pr(x) = 0.99974$$

We can see that the probability of accepting this shipment is very close to unity. If 5 defectives out of 100 units are more defectives than we are willing to live with, the plan is insufficient to determine the poor quality level of the batch. We must come up with a stricter plan.

In this solution, we had to assume the number of defectives in the shipment. But this is what we are trying to determine! How can we assume it? In general, in choosing a sampling plan, we must calculate P_a for a variety of possible defective levels and see if the plan behaves as we desire. We choose a plan based on its ability to accept defective levels we can live with and reject defective levels that we cannot tolerate.

The second part of the question is easy to answer. The average sample burden of any single sample plan is equal to the sample size n—in this case, 10. No matter what the quality of the incoming shipment, we will inspect 10 units. This would not be true in a double or multiple sampling plan, which would require a more probabilistic approach to determine the ASN. The interested reader is referred to the texts listed at the end of this chapter.

We have seen how an acceptance sampling plan is analyzed, but we have not discussed how such a plan is determined. That is a complex process and beyond the scope of this text. The basic process is one of trial and error, where a plan is assumed, analyzed, and modified until it meets the P_a versus quality that is desired. Alternatively, tables exist that can be used to suggest possible plans that will meet certain criteria.

V. SUMMARY

This chapter addressed maintaining designed-in levels of product reliability, safety, and quality. The focus was on evaluating and monitoring the manufacturing system (and the equipment therein) to ensure that it is capable of providing the optimal product from a benefit-cost perspective—that is, while still maintaining and providing the market-required levels of reliability, safety, and quality.

Reliability control focuses on maintaining the inherent level of product reliability that has been designed in and predicted. The objective is to integrate into the manufacturing system and process design specialized screens and supporting tests and inspections that eliminate, or minimize, product latent defects moving to the marketplace. Latent defects are not readily observable defects (or partial defects) that reveal themselves under a nonoverstress situation. The removal of these defects enables a product to move quickly out of the infant mortality phase and into the useful life phase of a product's characteristic life-cycle.

Safety control addresses both the product and the manufacturing system design. Designs must exhibit safety features that illustrate their benefits in both paper analyses and special tests. Analyses such as failure mode, effects, and criticality analysis (FMECA) and fault tree analysis (FTA) are excellent methods for assessing product and system safety cost-effectively. For the manufacturing system, sound design must be supported by a well-founded occupational safety program that focuses on human factors.

Quality control focuses primarily on patent defects; that is, defects that are readily observable. Once a product is designed with the customer's quality requirements in mind, it is up to the system to deliver consistently the corresponding product. Many statistical techniques are used to monitor product quality, including control charting as part of statistical process control (SPC).

VI. QUESTIONS

1. Discuss the relationship between design of experiment and reducing the product development cycle.
2. How are "life" and "MTBF" different?
3. How does durability fit into the concept of the reliability bathtub curve?
4. Derive the reciprocal relationship between failure rate and MTBF.
5. Discuss the difference between patent and latent defects.
6. Is reliability still an issue in space satellites even with the maintenance capability provided by the U.S. Shuttle aircraft?
7. When is it best to use inductive versus deductive analysis methods?
8. Discuss the legal liabilities associated with a lack of human factor design consideration.
9. What is the difference between process monitoring and acceptance sampling? Suggest several situations in which one would be more appropriate than the other.
10. List the advantages and disadvantages of single sampling plans versus double or multiple sampling plans. Think of a situation where each would be most appropriate.
11. Think of situations in your organization that might cause each of the following trends in a control chart: cyclic variation, gradual drift in the mean, a sudden jump in the mean, or two distinct populations.

VII. REFERENCES

1. Hicks, C.: *Fundamental Concepts in the Design of Experiments*. CBS College Publishing, New York, 1982.
2. Box, G., Hunter, W., and Hunter, J.: *Statistics For Experimenters*. John Wiley and Sons, New York, 1978.
3. Miller, I. and Freund, J.: *Probability and Statistics for Engineers*. Prentice-Hall, Englewood Cliffs, New Jersey, 1985.
4. Brauer, D.: "Materials And Reliability," *Course Report For University of Illinois at Chicago*. Chicago, 1986.
5. Kapur, K. and Lamberson, L.: *Reliability in Engineering Design*. Wiley, New York, 1977.
6. O'Connor, P.: *Practical Reliability Engineering*. Wiley, New York, 1985.
7. U.S. Department of Defense: *MIL-HDBK-781, Reliability Test Methods, Plans, and Environments for Engineering Development, Qualification, and Production*. 1984.
8. U.S. Department of Defense: *DoD-HDBK-344, Environmental Stress Screening of Electronic Equipment*, 1986.
9. U.S. Naval Avionics Center: *R&M-STD-R00217, Human Reliability Analysis*. 1986.
10. Anderson, R. and Lakner, A.: *Reliability Engineering for Nuclear and Other High Technology Equipment: A Practical Guide*. Elsevier, England, 1986.
11. U.S. Nuclear Regulatory Commission: *NUREG 0492, Fault Tree Handbook*. 1981.
12. U.S. Department of Defense: *MIL-STD-1629A, Procedures for Performing a Failure Mode, Effects, and Criticality Analysis*. 1980.
13. Nicholson, A. and Ridd, J. (editors): *Health, Safety, and Ergonomics*. Butterworths, England, 1987.
14. U.S. Department of Defense: *Human Engineering Design Criteria for Military Systems, Equipment, and Facilities*. 1989.
15. Lapin, L.: *Probability And Statistics For Modern Engineering*. Brooks/Cole, Monterey, California, 1983.
16. Banks, Jerry: *Principles of Quality Control*. John Wiley and Sons, New York, 1989.
17. Besterfield, Dale H.: *Quality Control*, 2nd ed. Prentice-Hall, Englewood Cliffs, New Jersey, 1986.
18. Taguchi, G., Elsayed, A., and Hsiang, T.: *Quality Engineering in Production Systems*. McGraw-Hill, New York, 1989.

System Maintenance

As an essential part of TMA, maintenance plays an important role in product quality, as well as in manufacturing system throughput. Inevitably, all manufacturing system equipment will degrade, and will fail eventually if allowed to do so. Since it is the system equipment that enables a process to be performed, system maintenance issues have great impact on the performance capability of a process. It is important to establish a cost-effective preventive maintenance program to enhance manufacturing system uptime and extend the useful life of expensive equipment and tooling.

This chapter provides several sections focusing on maintenance. The first section addresses maintenance program planning. The second discusses maintenance strategy, including the reliability-centered maintenace approach for classifying maintenance tasks as on-condition maintenance, condition monitoring, or hard-time maintenance in order to ensure cost-effectiveness. The final section presents maintainability analysis as a key element in developing the overall manufacturing system maintenance plan.

I. INTRODUCTION

A key element of all manufacturing systems is maintenance. Obviously, if a system received no maintenance it would run until failure. If we lived in a perfect world there would be no failure and our manufacturing systems would run forever. Our perfect world does not exist, however, and we are destined to experience system failures.

In working toward TMA, system maintenance is very important and should not be overlooked. When failures do occur, their cost can be very high, both in lost manufacturing time and in equipment repair or replacement. To minimize these costs it is necessary to have a comprehensive and cost-effective maintenance program.

The key to a successful program is to minimize total maintenance life-cycle costs. This is achieved by performing maintenance at the optimal time between maintenance (TBM) interval for each major system component. The curves presented in Figure 1 illustrate this concept. The optimal maintenance interval is achieved by balancing the performance of corrective maintenance (CM) and preventive maintenance (PM).

The figure illustrates that the cost of CM increases with time. This is because cost increases with the severity of the problem. By performing PM, costly corrective repairs can be avoided. But PM can be expensive too, and it is desirable to extend the time between its performance, not only for cost reasons but also to avoid introducing system problems caused by too much hands-on interaction.

This chapter provides insight into establishing a cost-effective maintenance plan for the manufacturing system. Special attention is given to establishing a maintenance program that cost-effectively enhances system availability. This includes addressing the fundamental ideas of maintenance planning. Emphasis is also placed on establishing a maintainability program and deriving quantitative data as part of enhancing the cost-effective achievement of maintenance.

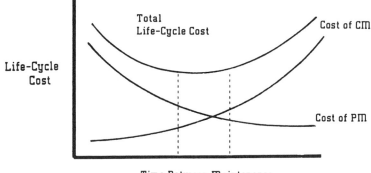

Figure 1   Maintenance Life-Cycle Cost Curve

II. MAINTENANCE PROGRAM PLANNING

A. Types of Maintenance

The performance of maintenance serves two functions: (1) correct failures, and (2) delay the onset of failures. Preventive maintenance provides the delaying action, and corrective maintenance provides the corrective action. For either category of maintenance, the focus is on wearout failures that occur as a result of material wear, fatigue, corrosion, and so on.

Underlying maintenace theory is the concept of *force of mortality*. It is represented by the hazard rate, or instantaneous failure rate. The force of mortality causes the hazard rate to increase. It is through our maintenance program that some control can levied against the force of mortality.

In Chapter 6, we stated that it is important to quickly move products away from infant mortality and have them operate in chance mortality, or useful-life. In this chapter, we look at the other end of the reliability "bathtub." It is important to keep manufacturing systems away from wearout mortality and have them operate in their useful-life mortality phase.

In essence, manufacturing systems are products themselves, and the same mortality concepts apply. The goal for both is to remain in the useful-life phase of the life cycle.

In discussing maintenance, the curves illustrated in Figure 2 are key. As shown, over time the hazard rate increases. Ultimately, all systems and equipment move into wearout if they receive no special attention. The objective of maintenance is to pull an item out of the wearout phase or, in the case of PM, to prevent or delay its entrance into the wearout phase.

As a system ages, surviving t increments of time, the probability that it will fail during the next increment increases. This probability represents the force of mortality. Thus, it becomes clear that maintenance concerns itself with phenomena that occur as a result of aging.

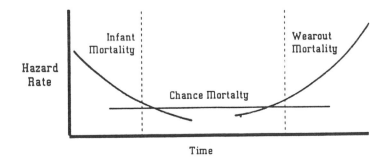

Figure 2 Force of Mortality Curves

It is desirable to avoid wearout failures and to experience only chance (or useful-life) failures. With chance failures the age of a system is of no consequence. Therefore, a system is considered like new as long as it is still operating and has not reached its wearout phase. During this period, the force of mortality is constant.

Through a comprehensive PM program, it is possible to keep a system operating in a state where the force of mortality is constant. In other words, the manufacturing system remains in its useful-life period and exhibits only chance, or random, failures.

The significance of this maintenance approach is illustrated in Figure 3. Consider the hypothetical case where "perfect" CM is possible. The hazard rate curve for perfect maintenance serves as a dividing line between PM and CM. Note that the hazard rate becomes constant with perfect CM. This indicates that wearout failures can also occur by chance [1]. This chance characteristic is caused by the unequal age of the components. As a result, wearout failures cannot be distinguished from pure chance failures.

Perfect CM implies that replaced or repaired items work properly. Also, no damage or error occurs during the repair task. However, perfect CM does not exist. The hazard rate for faulty CM lies above that which would result from perfect CM. In essence, the item is being pulled back into its useful-life phase.

Therefore, PM plays a critical role in maintaining a low hazard rate. In PM, items are replaced or repaired before the time a malfunction might occur. Consequently, the hazard rate will lie below that of the hazard rate resulting from perfect CM.

Also depicted at the bottom of Figure 3 is the constant hazard rate resulting from pure chance failures. It is important to realize that these failures cannot be reduced by maintenance. However, they can be increased by careless or faulty maintenance.

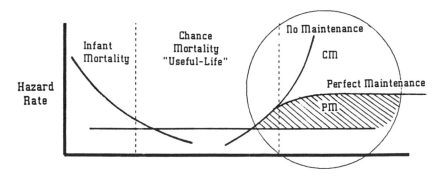

Figure 3   PM vs CM

The maintenance strategy developed for the manufacturing system and its components identifies the type of maintenance to be performed. A key issue in developing the maintenance strategy is cost. Reliability-centered maintenance (to be discussed later in this chapter) is an engineering method for determining the most cost-effective type of maintenance.

Two cost factors exist in maintenance: (1) the actual expenditure for maintenance action, and (2) the gain obtained by maintenance action. These two factors must be balanced if we are to make optimal maintenance decisions. For example, low replacement cost of a failed item may rule out the use of PM. On the other hand, high replacement cost may necessitate PM.

B. The Planning Process

Implementation of PM and CM requires the development of a detailed maintenance plan. A good maintenance plan provides an organized and disciplined approached to ensuring a high level of manufacturing system availability. It also ensures that the system operates efficiently safely, as intended.

The scope of a sound maintenance program plan includes the overall manufacturing system and its interaction with other ongoing activities, particularly production scheduling, However, the plan must provide the detail necessary to properly maintain each key maintenance item (KMI), and parts thereof, comprising the system.

Figure 4 illustrates the process for developing a maintenance program plan. In general, there are six fundamental steps [2].

The maintenance plan defines the requirements and tasks necessary to restore or sustain the operating capability of the system. It evolves from various analyses to identify the key elements of the program. These elements include

1. Maintenance strategy
2. Reliability, maintainability, and availability requirements
3. Specific maintenance tasks
4. Maintenance organizations
5. Support and test equipment requirements
6. Maintenance standards
7. Supply support requirements
8. Facility requirements
9. Technical publications

The foundation of the overall program plan is management. The focus is on the assignment of maintenance responsibilities. With this defined, it becomes possible to prepare the detailed maintenance schedule for each KMI. These schedules control the performance of all known maintenance tasks in accordance with established manufacturing system and production priorities.

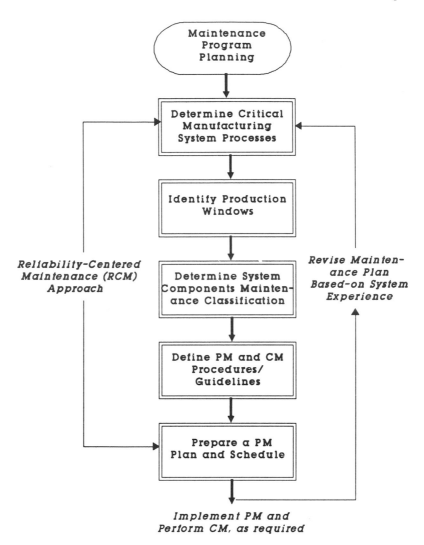

Figure 4   Maintenance Program Planning Process

Maintenance schedules are relatively firm for PM activities accomplished on a periodic basis. Adjustments can be made to compensate for variations caused by operational requirements and current workloads. The PM schedules themselves are developed based on system reliability data, actual or estimated, as well as the best judgment of material maintenance specialists. Defining PM intervals will be addressed later in this chapter.

CM cannot be scheduled in the same sense as can PM. The reason is that random failures occur despite the performance of PM. Once a failure occurs, the appropriate CM task must be scheduled into daily activities. Remember that the need for CM can arise at any moment. Therefore, the maintenance program plan must enable the responsible maintenance organization to respond in a timely manner.

1. Maintenance-Induced Problems

We previously touched on how a broad range of factors contribute to the degradation of manufacturing system availability. This degradation is usually the result of the interaction of human, machine, and environment. This is particularly true during system maintenance.

Manufacturing systems are defenseless against poor maintenance practices. Excessive handling caused by frequent and/or faulty maintenance often results in defects being introduced into the system. Typical examples include foreign objects left in an assembly, bolts improperly torqued, dirt injection, parts replaced improperly, wrong parts used, and improper lubricants.

Although it is important that a system be highly maintainable, maintenance tasks should be designed out where practical. But keep in mind that even with these efforts, poorly trained, poorly supported, or poorly motivated maintenance technicians can have a catastrophic effect on system availability.

2. Maintenance Documentation

A good way to offset maintenance-induced problems is to have sound maintenance documentation available. This includes written, graphical, and other types of information to guide maintenance technicians in accomplishing both the PM and CM tasks identified in the system maintenance plan.

All maintenance documentation reflects the overall maintenance strategy and repair policies established for the system. A comprehensive documentation package provides clear-cut direction leading from failure detection and fault isolation to the actual repair task. Also, maintenance procedures must be clear and presented in a format that is easy to understand and update.

Typically, four types of information represent the minimum package for ensuring that the manufacturing system is successfully operated and maintained. It is critical that these documents remain dynamic, clearly written, and up-to-date.

The four information types include

1. Functional description and operating instructions for each major system component
   a. A description of system capabilities and limitations
   b. A technical description of the system operation

    c. Step-by-step operating procedures
    d. Confidence checks to verify satisfactory system performance
2. Equipment and installation description
    a. Flow diagrams
    b. Schematics
    c. Parts data in sufficient detail to permit reordering or fabrication
    d. Instructions for installing and checking out installed or retrofitted components
3. Maintenance aids (for troubleshooting)
    a. Methods for system-level fault isolation when the system is up but operating in a degraded mode (use and interpretation of system readiness test results)
    b. Methods for system-level fault isolation when the system is totally down (use and interpretation of fault isolation test and monitor console displays)
    c. Procedures for functional equipment-level fault isolation (based on fault-sensing indicators supplemented, as required, by test point measurements using built-in test equipment)
    d. Equipment-level isolation techniques to permit identification of the problem area within a single module or replaceable part (for example, maintenance expert systems)
    e. Routine test, adjustments, alignment, and other preventive procedures performed at periodic intervals
    f. Reference to failure mode, effects, and criticality analysis (FMECA)
4. Ready reference documentation (information that is routinely required by the maintenance technician and is easily usable in the work area
    a. Routine checkout, alignment, and PM procedures
    b. Fault-monitoring interpretation and replacement data
    c. Supplemental troubleshooting techniques required to complement the automatic fault detection and isolation system
    d. Item and unit spare parts ordering data keyed to system identity codes

III. MAINTENANCE STRATEGY

A.   Reliability-Centered Maintenance

The cornerstone of an effective maintenance plan is the maintenance strategy. The maintenance strategy defines the application PM and CM for each major component, and parts therein, of the manufacturing system. By extending the overall strategy to the system's KMI indenture level, it is possible to develop an optimal maintenance plan.

An engineering method for defining the KMI-level maintenance strategy is reliability-centered maintenance (RCM). The RCM method uses a decision logic for systematic analysis of failure mode, rate, and criticality data. This method enables creation of the most effective maintenance requirements for KMIs. As a result, the scheduled maintenance burden and support costs are reduced while system availability is sustained.

The RCM method focuses engineering attention on the major system or equipment in a disciplined manner. Benefits include

1. Development of high-quality maintenance plans in less time and at lower cost
2. Availability of a maintenance history for the system and the KMIs therein
3. Assurance that all KMIs and their critical failure modes are considered in the maintenance requirements
4. Increased probability of optimal requirements
5. Online information exchange among engineering and management staff

The RCM method is first applied during the system design and development phase. It is then reapplied, as appropriate, to sustain an optimal maintenance program based on actual operating experience.

RCM segregates KMIs into two distinct categories: (1) non-safety-critical components, and (2) safety-critical components. For non-safety-critical components, scheduled maintenance is performed only when the task will reduce the life-cycle cost of ownership. For safety-critical components, scheduled maintenance is performed both when the task will prevent a decrease in reliability and/or a deterioration of safety to an unacceptable level and when the task will reduce the life-cycle cost of ownership.

The fundamental premise of RCM is that reliability is a design characteristic to be achieved and preserved throughout a manufacturing system's operational life. Based on this premise, an RCM logistics support program can be developed using a decision logic that focuses on the consequences of failure [3,4]. We can then easily apply the RCM method to define a maintenance program that will provide the desired, or specified, levels of operational safety and reliability at the lowest possible overall cost.

The cost goal is achieved by targeting maintenance problem areas for the PM program. Fundamentally, PM consists of routine inspections and servicing. It is intended to correct and detect potential failure conditions and make corrections that will prevent major operating difficulties. It is most effective when service requirements are known or failures can be predicted with some accuracy.

PM is desirable when it increases the operating time of an item by reducing the severity and frequency of breakdowns. It typically includes cleaning, lubricating, inspection, calibration, testing, critical part replacement before failure, or complete overhauls.

When system failures occur, they idle workers and machines, result in lost production time, delay schedules, and require costly repairs. Therefore, performing CM may not be the best maintenance strategy. On the other hand, PM should be performed only when it provides a cost-benefit over CM. A cost trade-off exists, and PM can be carried too far. When immediate repair is not necessary and little harm is done by waiting, CM is advantageous.

Using the RCM method results in defining a cost-effective PM program where

- Incipient failures are detected and corrected either before they occur or before they develop into major problems.
- The probability of failure is reduced.
- Hidden failures are detected.
- The cost-effectiveness of a maintenance program is improved.

B. The RCM Process

The RCM process aids in determining the specific maintenance tasks to be performed, as well as influencing system design maintainability and reliability. The process is based on a decision logic. Each KMI potential failure mode is evaluated using the decision logic to identify the maintenance tasks to be performed as part of the overall maintenance plan.

The process forces maintenance tasks to be classified into three areas:

1. Hard-time maintenance, for those failure modes that require scheduled maintenance at predetermined, fixed intervals of age or usage.
2. On-condition maintenance, for those failure modes that require scheduled inspections or tests designed to measure deterioration of an item. Based on an item's deterioration, either CM is performed or the item remains in service.
3. Condition monitoring, for those failure modes that require unscheduled tests or inspection of components whose failure can be tolerated during operation of the system, or where impending failure can be detected through routine monitoring during normal operation.

The following seven steps comprise the RCM method [5].

1. Step 1: Determine KMIs

This is done by performing a failure mode, effects, and criticality analysis (FMECA) and/or a fault tree analysis (FTA) (as discussed in Chapter 6). These analyses identify KMIs and provide quantitative failure mode data to help answer the RCM decision logic questions. This includes determining the effect of component malfunction, human error, and other potential failures on the system and their priority for improvement.

2. Step 2: Acquire Failure Data

Data is compiled for each basic fault identified during Step 1. Part failure rate, operator error, and inspection efficiency data are necessary inputs for determining occurrence probabilities and assessing criticality.

Part failure rate data is typically available in either in-house or industry-wide data systems. Operator error is represented by the probability that a failure caused by an operator will take place, whether intentionally or unintentionally. This data is developed through subjective techniques based on discussions with persons familiar with the system operating environment. Inspection efficiency is the probability that a given defect will be detected before leading to a failure. It is treated similarly to operator error probability.

3. Step 3: Develop Criticality Data

This involves computing the criticality probability of each potential failure mode. Sensitivity, or conditional probability, is the probability that a defined manufacturing incident will occur given that a basic system fault has occurred. The sensitivities are then used to compute the criticality of each basic fault. Criticality is the measure of the relative seriousness or effect of each fault on the top event. It involves both qualitative engineering analysis (for example, visual analysis of FTA minimal cut sets) and quantitative analysis (for example, exercising data through the fault tree). This provides a basis for ranking the faults in order of severity. Criticality is defined by

$$Cr = P[X_i] \times P[FX_i]$$

where $P(FX_i)$ is the sensitivity and $P(X_i)$ is the occurrence probability of the basic fault.

4. Step 4: Apply Decision Logic to Critical Failure Modes

This involves using the RCM decision logic to define the most effective PM task combinations for each KMI. As each failure mode is processed, judgements are made about the necessity of various maintenance tasks. The selected tasks, together with the performance intervals deemed appropriate, define the total scheduled PM program.

The decision logic is designed in two levels. Level 1 (Questions 1-4, see Figure 5) requires evaluation of each failure mode for determination of the consequence category. Level 2 (Questions 5-26; see Figures 6 and 7) takes the failure mode(s) into account for selecting the specific type of task.

The Level 1 questions lead in to Level 2 questions via one of the following consequence categories:

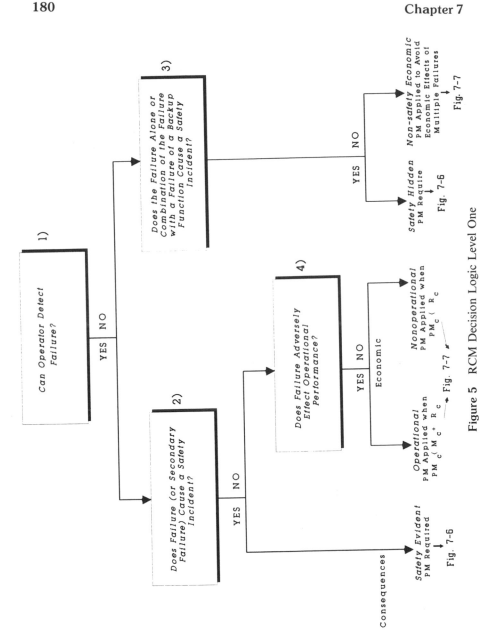

Figure 5   RCM Decision Logic Level One

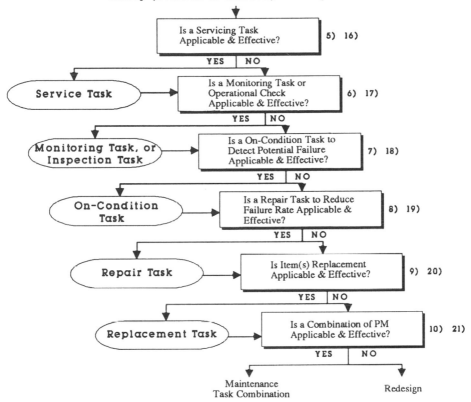

Safety (Evident or Hidden) Consequence

Is a Servicing Task Applicable & Effective? — 5) 16)

YES | NO

Service Task

Is a Monitoring Task or Operational Check Applicable & Effective? — 6) 17)

YES | NO

Monitoring Task, or Inspection Task

Is a On-Condition Task to Detect Potential Failure Applicable & Effective? — 7) 18)

YES | NO

On-Condition Task

Is a Repair Task to Reduce Failure Rate Applicable & Effective? — 8) 19)

YES | NO

Repair Task

Is Item(s) Replacement Applicable & Effective? — 9) 20)

YES | NO

Replacement Task

Is a Combination of PM Applicable & Effective? — 10) 21)

YES | NO

Maintenance Task Combination | Redesign

Figure 6   RCM Decision Logic Level Two

1. Safety, Evident (Questions 5-10)—PM tasks are required to ensure safe operation. All questions must be asked. If no effective task results from this category analysis, then component redesign is mandatory.

2. Economic, Operational (Questions 11-15)—A PM task is desirable if its cost is less than the combined cost of the operation loss and the cost of repair. Analysis of the failure causes through the logic requires the first question (Servicing) to be answered. Either a Yes or a No answer to Question 11 still requires movement to the next level. From this point on, a Yes answer completes the analysis and the resultant task(s) will satisfy the requirements. If all answers are No, then no task has been generated and if economic penalties are severe, a component redesign may be desirable.

Non-safety or Economic (Operational/Nonoperational) Consequence

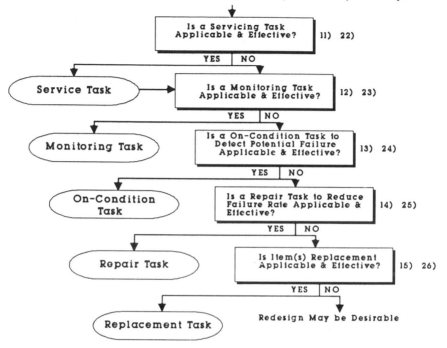

Figure 7 RCM Decision Logic Level Two

3. Economic, Nonoperational (Questions 11-15)—A PM task is desirable
    if its cost is less than the cost of repair. Analysis of failure causes is the
    same as Economic, Operational.
4. Safety, Hidden (Questions 16-21)—PM tasks are required to ensure the
    availability necessary to avoid the safety effects of multiple failures.
    All questions must be asked. If no tasks are found to be effective, then
    redesign is mandatory.
5. Nonsafety, Economic (Questions 22-26)—PM tasks are desirable to
    ensure the availability necessary to avoid the economic effects of multi-
    ple failures. Analysis of failure causes is the same as Economic, Opera-
    tional.

   During a decision logic application, at the user's option, advancement to
subsequent questions is allowable after a Yes answer—but only until the cost
of the last task being considered is equal to the cost of the failure prevented.
   Notice that default logic is reflected in the economic consequence categories
by the arrangement of the logic questions. In the absence of adequate informa-

tion to answer Yes or No to questions in the second level, default logic dictates that a No answer be given and the subsequent question be asked. As No answers are generated, the only choice available is the next question, which in most cases provides a more conservative and costly route.

Developing the most effective maintenance task strategy is handled similarly for each consequence category. To determine the most effective maintenance task, it is necessary to process each failure mode through Level 2 of the logic diagram.

For the economic, operational and nonoperational failure consequence categories, it is necessary to make a cost-effectiveness decision before moving into Level 2 of the logic. If the expected cost of system failure per period without PM is greater than the expected cost of system failure with PM, then PM is the best strategy.

The expected cost of system failure per period, if there is no PM, is the cost of system failure divided by the expected number of periods between system failures [6].

$$TC = C_b/E(n)$$

where: $C_b = Nc_b$ is the total cost of system failure

N is the total number of items in a group

c_b is the cost of a single failure

$E(n) = \sum_n nP_n$ is the expected number of periods between failures

$1/E(n)$ is the expected number of failures per period

n is the time period

P is the probability of a breakdown in period n

The expected cost of system failure per period with PM includes both the cost of the PM and the cost of those components that fail regardless of PM.

$$TC = [C_{pm} + (B_n \times C_b)]/n$$

where: $B_n = N(P_1 + P_2 + \cdots + P_n) + B_1 \times P_{n-1} + B_2 \times P_{n-2} + \cdots + B_{n-1} \times P_1$

C_{pm} is the cost of PM

n is the number of time periods between PM

B_n is the expected number of failures with PM performed every n time periods

PM is more economical than CM if

$$[C_{pm} + (B_n \times c_b)]/n > C_b/E(n)$$

Notice that if standby capacity exists in the system, a single system failure may not be critical. (This assumes that the manufacturing process can be performed elsewhere in the overall manufacturing system.) Excess capacity favors CM over PM. When system utilization approaches capacity, PM is more desirable.

5. Step 5: Compile/Record Maintenance Classifications

Applying the decision logic (in Step 4) segregates maintenance requirements into the three general classifications noted at the beginning of this section [7].

The maintenance-task profile indicates, by part number and failure mode, the PM strategy task selection for RCM logic questions that were answered Yes.

6. Step 6: Implement RCM Decisions

With the maintenance-task profile established, the task frequencies/intervals are set and integrated into the overall maintenance program plan. Setting the task frequencies/intervals requires first determining whether applicable data are available that suggest an effective interval for task accomplishment. Prior knowledge is used, as applicable, to define a scheduled maintenance task that will be effective and economically worthwhile.

If there is no prior knowledge, the task frequencies/intervals are established initially by experienced engineering and maintenance personnel. Good judgment and actual operating experience are used in concert with accurate reliability data. If failures are adequately modeled by a Poisson process, then the Poisson distribution can be used to predict the probability of occurrence of failures in a given period of time. With a target reliability and failure rate known, a PM interval can be calculated for each part.

7. Step 7: Apply Sustaining Engineering Based on Actual Experience Data

The RCM process has a life-cycle perspective. The driving force is reduction of the scheduled maintenance burden and support cost while maintaining the necessary manufacturing readiness state. Therefore, it is prudent to review the RCM information, as available maintenance and reliability data move from a predicted state to actual operational experience values.

IV. MAINTAINABILITY ANALYSIS

Maintainability analysis is performed to determine if a manufacturing system that satisfies its operational requirements can be maintained cost-effectively. The analysis serves four main purposes: (1) to establish design criteria; (2) to allow for design evolution based on trade-off studies; (3) to contribute toward defining maintenance, repair, and servicing policies; and (4) to verify design compliance with the defined maintainability requirements.

Maintainability itself is a measure of the ease and speed with which the system can be restored to operational status following a failure. It is represented by the parameter mean time to repair (MTTR), which is a measurable and controllable parameter that is specified during design, measured during test, and sustained during operation. The achievement of a high level of maintainability is primarily a function of system design.

An integral part of a maintainability analysis is the performance of a maintainability prediction. Such a prediction supports design and development by deriving repair time, maintenance frequency per operating hour, PM time, and other parameters.

The calculation of MTTR is based on the active repair time associated with the four time elements of fault isolation, fault correction, calibration, and checkout (see Figure 8). Those time elements related to preparation and delay, while quantifiable, do not provide much insight into the maintainability design of manufacturing equipment.

A. Maintainability Improvement

The maintainability of a manufacturing system (and the equipment therein) is highly dependent its design. Therefore, it is essential that the system be designed for ease of maintenance. This involves considering design features that address automatic detection, location, and diagnosis of failures, and the incorporation of easily accessible and interchangeable modules and subassemblies.

Additionally, the following design maintainability features are common:

1. Modular design techniques, such as uniform size and shape, guide pins, and keyed connectors, ease of test/checkout, quick disconnect, and minimum number of functions
2. Built-in fault-detection circuits
3. Design for replacement at higher levels
4. Design-level fault diagnosis localization ability
5. Built-in-test (BIT) capabilities
6. Readily accessible and identifiable test points
7. High-quality technical manuals and maintenance aids
8. Limited access barriers to replaceable items
9. Limited number of interconnections per replaceable item
10. Use of plug-in elements
11. Minimal requirements for special maintenance tools

There are also maintainability improvement techniques that do not directly impact quantitative parameters, such as:

1. Performance monitoring, the identification of performance degradation that may indicate impending system failure (thus preparing maintenance

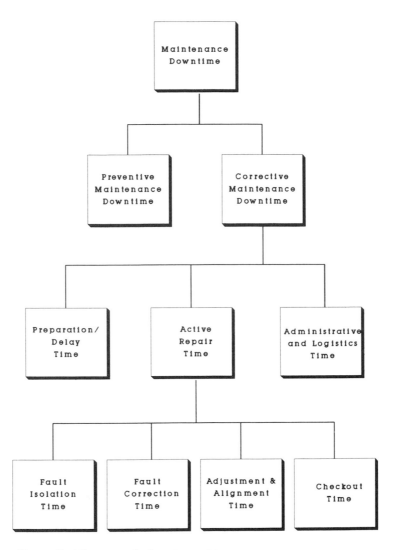

Figure 8   Elements of Time Comprising Maintenance Downtime

personnel for the type of repair task to be encountered) or identify an
existing fault
2. Personnel training, the level and expertise of maintenance personnel
    trained for the proper performance of maintenance tasks
3. Levels of support (organizational/intermediate/depot), the maintenance
    philosophy prescribed for repair of failed hardware

4.   Spares support, the levels of spare support for failed units

B.   Automated Fault Isolation

In many cases, it is advantageous to incorporate some degree of automation in performance monitoring and fault isolation. This involves the integration of BIT features and automated test equipment (ATE) into the manufacturing system. ATE provides the following functions:

- Automatic control of test sequence and the selection of appropriate stimuli
- Comparison of monitored responses with predetermined standards
- Display or recording of test results

Automatic fault isolation functions are not a substitute for good maintainability design. Automatic testing is costly, and if it is not properly integrated into the system, it can induce more reliability and maintainability problems than it solves. When properly used, however, automatic testing reduces CM time and increases system availability.

In some cases, the automatic test features are used to detect (or predict) impending failure. This permits system degradation problems to be corrected as a PM routine, thereby increasing reliability. Automatic fault isolation techniques also reduce both the number of maintenance personnel and the maintenance skill levels required for the system.

Selection of the test features involves consideration of the following requirements and constraints:

1.   Test function
2.   Test modes
3.   Level of detection
4.   Degree of fault isolation

System-level design requirements for automatic fault isolation features must be (to the extent possible) specified qualitatively and defined quantitatively in the overall manufacturing system specification. This includes defining the degree of failure detectability; false alarm rates; degree of fault isolation to be provided; fail-safe provisions; and reliability and maintainability of sensors, interface hardware, and ATE.

C.   Maintainability Levels

When looking at manufacturing system maintainability, it is desirable to develop a maintainability functional level diagram (Figure 9). This is done by dividing the system into its various physical subdivisions, beginning with the highest and continuing down to parts that may be replaced, repaired, or adjusted during CM. This facilitates the designation of levels at which repair

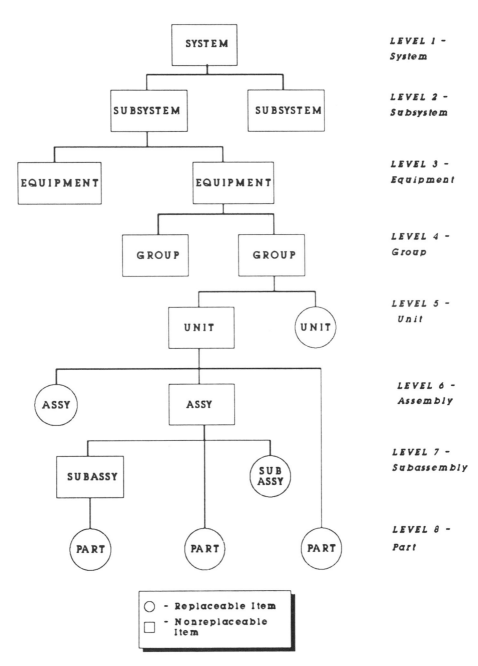

Figure 9   Functional Level Maintainability Diagram

is to be accomplished (either by modular replacement or by in-place repair) and the level to which fault isolation is to be extended.

D. Maintainability Prediction

As stated earlier, a maintainability prediction provides a quantitative evaluation of the manufacturing system (and equipment therein) in terms of mean time to repair, as well as other parameters. These parameters indicate the capability of the system to meet all specified quantitative maintainability requirements, including MTTR objectives and allocations.

The prediction method is applicable to any operational environment and type of system (that is, electronic or mechanical). It enables one to monitor the overall system maintainability throughout design and development. Thus, if it appears that the maintainability requirements are in danger of not being met, system design changes can be made before they become prohibitively expensive.

The prediction method is applied initially by making use of gross design data. It is revised as the manufacturing system design evolves. The method is applicable to any system level and is based on the performance of CM.

CM actions consist of preparation, fault isolation, and fault correction. Fault correction is further reduced into disassembly, interchange, reassembly, alignment, and checkout. The time to perform each of these tasks is an element of MTTR.

These elements are defined as follows:

Preparation—Time associated with those tasks that must be performed before fault isolation can be executed
Isolation—Time associated with those tasks required to isolate the fault to the level at which fault correction begins
Disassembly—Time associated with gaining access to the replaceable item or items identified during the fault isolation process
Interchange—Time associated with the removal and replacement of a faulty replaceable item or suspected faulty item
Reassembly—Time associated with closing up the equipment after interchange is performed
Alignment—Time associated with aligning the system or replaceable item after a fault has been corrected
Checkout—Time associated with verifying that a fault has been corrected and the system is operational

The time dependency between the probability of repair and the time allocated for repair typically produces a probability density function in one of three common forms: (1) normal, (2) exponential, or (3) lognormal. The

Normal Distribution applies to the relatively straightforward maintenance tasks and repair actions that consistently require a fixed amount of time to complete. The use of automated fault isolation techniques make this choice of distributions most appropriate.

Historically, the Lognormal Distribution was used to represent maintenance task performance, because of the excessive amount of time required to isolate the problem. However, in recent years, electronic design capabilities have moved the time emphasis to the actual repair task.

There are several quantitative parameters of interest in describing a manufacturing system's maintainability. These are defined in the remainder of this subsection [8].

1. Mean Individual Correction Maintenance Task Time, $\overline{M}_{cti}$. The mean time required to complete an individual maintenance task or action is

$$\overline{M}_{cti} = (\Sigma \, M_{cti})/N$$

where M_{cti} is the CM time required to complete the ith individual maintenance task or action and N is the number of observations.

2. Mean Time to Repair, $\overline{M}_{ct}$ or MTTR. The mean time required to complete a maintenance action over a given period of time is

$$\overline{M}_{ct} = \Sigma \, (\lambda_i \times \overline{M}_{cti})/\Sigma \, \lambda_i$$

where λ_i is the failure rate of the ith element of the item for which the prediction is being performed.

3. Mean Preventive Maintenance Time, $\overline{M}_{pt}$. The mean system downtime required to perform scheduled PM, excluding any PM time used during system operation and administrative and logistic downtime, is

$$\overline{M}_{pt} = \Sigma \, (f_i \times \overline{M}_{pti})/\Sigma \, f_i$$

where f_i is the frequency of individual (ith) PM action in actions per operating hour adjusted for system duty cycle and $\overline{M}_{pti}$ is the average time required for the ith PM action.

4. Maintenance Downtime Rate, MDT. The maintenance downtime rate per operating hour is composed of downtime due to CM and downtime required for PM.

$$\overline{MDT}_{ct}$$

5. Corrective Downtime Rate, $\overline{MDT}_{ct}$. The CM downtime per hour of operation is

$$\overline{MDT}_{ct} = \Sigma \, (\lambda_i \times \overline{M}_{cti})$$

6. Preventive Downtime Rate, $\overline{MDT}_{pt}$. The PM downtime per hour of operation is

$$\overline{MDT}_{pt} = \Sigma \, (\lambda_i \times \overline{M}_{pti})$$

7. Total Downtime Rate, $\overline{MDT}$. The total downtime for CM and PM combined is

$$\overline{MDT} = \overline{MDT}_{ct} + \overline{MDT}_{pt}$$

As indicated throughout this chapter, a key requirement for manufacturing systems is a high level of operational readiness or availability. Availability is the probability of being operationally ready at any point in time. It is similar to reliability, except that it takes into account system maintenance.

Operational availability is derived from

$$A_o = \text{MTBF}/(\text{MTBF} + \overline{MDT})$$

where MTBF is the mean time between failure (see Chapter 6) and $\overline{MDT}$ is the mean downtime for maintenance, including active repair time, administrative time, and logistic delay time.

The "inherent" availability of the system reflects the characteristic of design without consideration of administrative or logistic time. It is derived from

$$A_i = \text{MTBF}/(\text{MTBF} + M_{ct}, \text{ or MTTR})$$

V. SUMMARY

This chapter addressed the importance of maintenance to the manufacturing system. The implementation of TMA depends on the ability of the manufacturing system to be up and running in a mode that makes products adhering to the customers' performance requirements. We want to maximize system availability; that is, make sure that the system's uptime is much greater than its downtime. Maintenance plays a key role in maximizing system maintenance.

One of the first things that must be done with respect to maintenance is to develop a program plan. There are two fundamental types of maintenance: (1) preventive maintenance (PM) and (2) corrective maintenance (CM). The program plan defines the maintenance strategy, which takes into account both PM and CM activities. The main difference between the two is that PM is scheduled, whereas CM cannot be scheduled ahead of time.

The reliability-centered maintenance (RCM) approach to defining specific PM tasks for key maintenance items (KMIs) enables the most cost-effective PM program to be established. The RCM approach uses a decision logic to lead the user to optimal PM tasks for the KMI. All tasks are classified as hard-time maintenance, on-condition monitoring, or condition monitoring.

Maintainability formulas are used to quantitatively assess the system design and operational performance level. These formulas are used to assess competing design trade-offs intended to enhance the ability to perform maintenance, as well as to track improvements in the overall maintenance program.

VI. QUESTIONS

1. Explain the maintenance life-cycle curve. What role does the time between maintenance play in accounting?
2. How does maintenance fit into a discussion about the reliability bathtub curve?
3. What is the primary difference between preventive maintenance and corrective maintenance?
4. What does preventive maintenance do in relation to the reliability bathtub curve?
5. Discuss the probability of perfect maintenance occurring.
6. What are the steps of the maintenance program planning process?
7. What is reliability-centered maintenance? What is its objective?
8. Define the three classifications of maintenance tasks.
9. Exercise the reliability-centered maintenance decision logic for a helicopter transmission.
10. What is the MTTR for products consisting of three parts each, having a failure rate of 10 failures/million hours, and having a repair time of 30 minutes?
11. What are the components of MTTR?
12. Discuss the advantages of automated monitoring capability.
13. Discuss the trade-offs that are present in the availability equation.

VII. REFERENCES

1. Pieruschka, E.: *Principles of Reliability*. Prentice-Hall, Englewood Cliffs, New Jersey, 1963.
2. Kelly, A.: *Maintenance Planning and Control*. Butterworths, England, 1987.
3. U.S. Department of Defense: *MIL-STD-1843, Reliability-Centered Maintenance for Aircraft, Engines, and Equipment*. 1985.
4. U.S. Department of Defense: *MIL-STD-1388-1, Logistic Support Analysis*. 1983.
5. Brauer, D. and Brauer, G.: Reliability-Centered Maintenance. *IEEE Transactions on Reliability*. 17-24, Vol. R-36, 1987.
6. Tersine, R.: *Production/Operations Management: Concepts, Structure, and Analysis*. North Holland, New York, 1981.
7. Brauer, D.: *NTIAC-85-1, Depot Maintenance Handbook*. Southwest Research Institute, San Antonio, Texas, 1988.
8. U.S. Naval Air Systems Command: *NAVAIR 01-1A-33, Maintainability Engineering Handbook*. 1977.

Part Four

SYSTEM IMPROVEMENT MONITORING

The goal of yesterday will be the starting point of tomorrow.

—Carlyle

8

Data System Planning

To know if something is working, failing, or improving, it is necessary to have an ongoing data-gathering effort. This is particularly true with TMA, because a corporation must be able to assess its relative success or failure to attain TMA.

This chapter addresses the fundamentals of establishing an in-house data system for capturing and evaluating manufacturing system and product reliability, safety, and quality data. Included are sections on data organization, structure, operation, and implementation of a TMA-focused data system. Such a data system is designed to capture quantitative and qualitative information and, most importantly, lessons learned, to support enhanced engineering management decisions.

I. INTRODUCTION

Assessing one's position with respect to attaining and maintaining TMA requires a useful data system to be established. A useful data system is one that compiles pertinent management and manufacturing system and product reliability, safety, and quality performance information. This type of data significantly enhances a corporation's ability to make strategically sound manufacturing management and engineering decisions.

Detailed planning is necessary to define and structure a data system, particularly with respect to the types and amount of data needed. We want to collect as much information as makes sense (that is, data that can and will be used intelligently—not too much data, but not too little). The breadth of the data system is a function of the costs and difficulty of gathering data, as well as of the costs associated with not collecting specific data.

Before actually implementing a TMA-focused data system, we need to define a few things, including (1) its purpose, scope, and usage; (2) its structure in terms of user responsiveness, both in the near term and as it evolves in the future; (3) the fundamental steps involved in its operation, as well as input sources; and (4) the various outputs and benefits to be derived. With these things accomplished, it is then possible to establish a working data system that can be accessed in real time.

This chapter provides an overview of the fundamentals of a TMA-focused data system. We emphasize the data system's role as an integral part of product engineering, manufacturing, and service activities. In addition, we will address corporate management data needs.

II. BENEFITS OF DATA ORGANIZATION

In Chapter 4, we discussed the importance of project management as part of the product development cycle. During this cycle, important aspects of project management are assessing the performance and effectiveness of a product development and manufacturing project and gathering and organizing data and information. This data comes from a wide range of tests, experiments, and inspections performed as an integral part of the effort. Also, supplementary (that is, generic) information deemed applicable to the case at hand may prove useful.

In general, much data is generated routinely through the evaluation and engineering of a specific product, as well as through subsequent related product development efforts. However, unless there is a way to capture the data, its usefulness is limited and valuable lessons learned are lost over time. It is in this light that the need for, and benefits of, a comprehensive data system are most easily recognized.

A good data system facilitates the early definition of essential parameter values (for design, manufacture, and support purposes) by its experienced-based content. The logical organization of a properly planned data system encourages a rapid and orderly design evolution toward a product meeting its reliability, safety, and quality (RSQ) requirements.

A TMA-focused data system (conceptualized in Figure 1) provides a way to determine essential cause and effect relationships. It represents a primary experience pool for planning new development projects and for improving existing manufacturing systems and products with respect to RSQ.

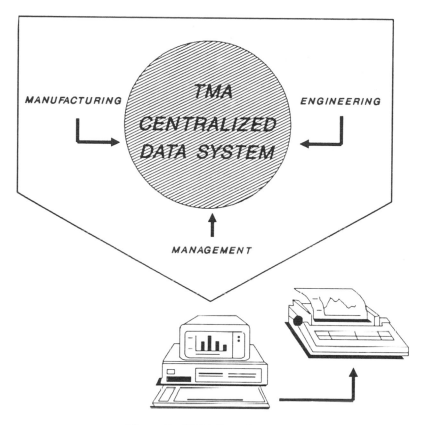

Figure 1   TMA Data System

To support the data system, a data recording and feedback program (to be discussed in Chapter 9) must be in place. This program is built around a formalized communication link between engineering, manufacturing, and product service, which allows for the efficient transmission of data and information inputs, as well as requests and inquiries. The key to making this program work effectively is to remove inhibiting departmental walls and communication barriers.

In conjunction with the data recording and feedback program, a data system typically performs the following functions:

1. Provides data resulting from ongoing activities for analyzing, evaluating, and assessing product manufacturability, performance, effectiveness, operational suitability, and logistic support requirements
2. Provides historical data that are applicable to the design and development of new products, processes, and experimental routines based on similar function, structure, or operational horizon

To perform these functions well, a data system also must satisfy management's need for current assessment information. In addition, it must incorporate the mechanics necessary for generalized data collection, storage, organization, processing, purging, and retrieval [1].

For each project undertaken, the data recording and feedback program identifies the specific data elements required. In addition to the types of data, the appropriate data recording format and media are determined. Early consideration of these factors not only helps to enhance the basic data system architecture, but also establishes the system's reporting burden and characteristics.

A primary focus of a TMA data system is on those data elements necessary to evaluate and verify the functional and operational characteristics of specific products and manufacturing systems and processes. But don't forget key data pertaining to day-to-day corporate management activities. The number of scope of potential TMA data elements can rapidly become overwhelming, particularly with the advent of automated documentation systems.

Table 1 provides a list (not exhaustive) of sample TMA data elements. The scope of a company's data system should reflect its specific needs. This includes a robust range of information and data to ensure effective and continuous movement toward TMA.

Although much of the basic data addressed in Table 1 is generated routinely, it is just as routinely not collected (due to lack of either initiative or management support). Consequently, it is lost as a resource. Having a working data system eliminates this problem by providing relevant data for current and future activities (for example, product development). It does this by providing insight and answers to questions that have been addressed previously, thereby eliminating the duplication of effort (for example, laboratory environmental testing). This characteristic enables a data system to acquire the confidence of all parties actively involved with the product throughout its life cycle.

III. DATA SYSTEM STRUCTURE

A centralized, comprehensive data system should collect and organize information such as inspection, test, failure, and repair data (and other related data). Such a data system provides maximum insight into design and operational deficiencies and enables the effectiveness of the various management and engineering activities to be tracked and measured. Such activities include design approaches, material and configuration alternatives, and improvements.

Perhaps the most difficult aspect of establishing a data system is implementing the mechanism (that is, the data recording and feedback program)

Table 1 TMA Data Elements

Management

- Operational requirements and performance measures
- System utilization (modes of operation and operating hours)
- System costs and effectiveness
- Operational availability, dependability, reliability, maintainability, safety, and quality
- Maintenance level and location
- Work-in-process tracking
- Payroll processing
- Personnel information
- Scheduling
- Line balancing
- Shop loading
- Equipment requirements
- Maintenance of engineering standards
- Piece-rate maintenance
- Integration with standard data systems
- Tracking operator responsibility
- Purchasing orders
- Sales literature
- Product data sheets
- Maintenance data sheets
- Parts catalogs
- Product schematics
- Technical documentation
- Product costs

Engineering

- Design reviews
- Value/cost analyses
- Modification releases
- Engineering releases
- Work order requests
- Material data sheets
- Setup sketches
- Operating instructions
- Tooling specifications
- NC operator sheets
- NC tape prove-cuts

Product Reliability

- Quantitative requirements
- Demonstration methods and results

(*continues*)

Table 1 (continued)

- Program milestones
- Procedures for evaluating and controlling design corrective action
- Reliability contributions to total design
- Supplier responsiveness
- Design review results

Product Safety

- Program monitoring
- Review, direction, and close-out of safety reports
- Procedure for initiating design correction action
- Failure mode and effects analysis results

Quality Assurance

- Quality assessment: destructive or nondestructive tests, 100 percent inspection, or statistical methods
- Frequency of inspection and defect levels
- Inspection records
- Metrology techniques and standards used
- Burn-in procedures
- Frequency of calibration of inspection equipment
- Procedures for configuration control

Product/Process Maintenance

- Spare/repair part types and quantities by location
- Supply responsiveness
- Item replacement rates

whereby the data and information can be captured; *not* generating data and information. For example, a product development/manufacturing effort is actually developing a technology base on relevant materials, system and component designs, performance characteristics, operational factors, and manufacturing techniques. A working data system formally classified and consolidates this engineering knowledge (including lessons learned) and then augments it with other similar product reliability, quality, maintenance, and safety information. This results in a significantly enhanced knowledge base that can materially reduce the time and cost of developing, manufacturing, and validating future products.

For example, in a case in which a company is involved in developing complex medical products and patient safety is of primary corporate concern, a key element of the data system must be reliability and safety information. The data system should be structured to highlight the effects of particular design features and interface stresses. This will provide valuable insight for product reliability and safety enhancement, as well as degradation control.

The real advantage of a TMA dedicated, centralized data system is its ability to provide detailed and specific data for use in various analytical exercises. As a matter of course, the data system catalogs and classifies all data and information in a form that makes it readily identifiable, accessible, and amenable to further analysis or evaluation.

Figure 2 illustrates a data system's structure which consists of a technical information base (TIB) and a quantitative data base (QDB) for each resident data bank. The TIB portion of the data bank accumulates product research and development reports, test reports, failure analysis and corrective action reports, design specifications, incident and field service reports, and other textual information concerning components and products. In addition to being used for engineering purposes, this information is frequently used to classify and interpret quantitative data.

The QDB portion of a data bank reflects relevant engineering parameters. The QDB includes functional performance, test results, maintenance, and so forth, addressed during product design, through manufacture, and into field operation.

Accessing this information could be a function of automated search and retrieval routines that key on selected descriptors and content identifiers. The selection of meaningful descriptors is essential in the development of a useful data system. Descriptors identify component reliability, quality, maintenance, and/or safety characteristics to enable later correlation among these characteristics and specific component and product design configurations, manufacturing efficiencies, and, ultimately, commercial applications.

An evolutionary approach is typically used in establishing data systems [2]. Ultimately, an automated system featuring real-time data entry and user

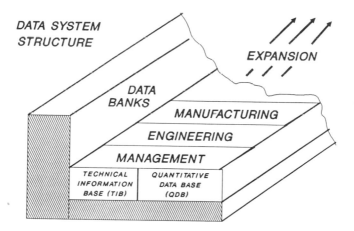

Figure 2　Data System Structure

access if desirable. Initially, a more modest approach may be appropriate, in which the system relies on manual entry and retrieval of data.

Because of the availability of low-cost personal computers and commercial data-base software, some initial automation may be practical. At first, output data and information might be transmitted via hard-copy or floppy-disk reports. Since networking is rapidly becoming economical, however, it probably could be adopted after the data system becomes fully operational. As the data system grows, a more powerful central computer or distributed processing capability may become practical.

As stated previously, the data system must be organized as a centralized entity providing one point of data management (that is, submittal and distribution). The data system also must be readily accessible to all potential users; that is, engineering, manufacturing, and product service. Also, the data must be accurate. Use of the system should be encouraged through special alerts, narrative reports, and other outputs. These outputs notify involved parties about specific problems, trends, and corrective actions.

In general, it is wise to publicize all uses of the data. Additionally, where practical, the collection of data and information should be compatible with that of other data systems accessible by all intended users, whether industry-wide or internal.

IV. DATA SYSTEM OPERATION

The general operational procedure for compiling and reducing applicable engineering, manufacturing, and field (service) experience data consists of three tasks: (1) data collection, (2) data processing, and (3) data reporting. As part of the data system's operational protocol, each task must be flexible enough to adapt to the specific input material and properly process and output a desired range of information.

Figure 3 depicts the operational cycle of the data system. As part of this cycle, four distinct milestones are defined: (1) data recording, (2) data transmission, (3) data storage, and (4) data retrieval. Initially, all appropriate data is recorded (Milestone 1) in accordance with the overall data recording and feedback program that is in place. Data and information are recorded on approved corporate forms designed for efficient, yet comprehensive, capture of pertinent facts.

A key source of data is a uniform and coordinated mechanism for reporting, analyzing, and implementing corrective actions for manufacturing system and product failures. This includes those failures that occur during laboratory qualification and reliability tests, manufacturing tests, and field operation and repair activities. Capturing this data requires that failures be accu-

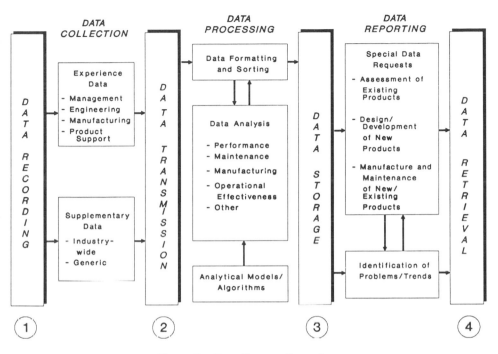

Figure 3   Data System Operation

rately reported and thoroughly analyzed (and that corrective actions be taken on timely basis to reduce or prevent recurrence).

Once recorded, all data and information are transmitted (Milestone 2) to the data system. At this point, the material is formatted and sorted into the applicable data bank(s). Once in the data bank(s), the data undergoes analysis by appropriate analytical methods that combine the new data, data already resident in the data bank, and the overall data system. After the analytical routines are run, the data is stored (Milestone 3) as part of permanent memory for future reference.

The data and information resident in the data system are retrievable (Milestone 4) on an as-needed basis (that is, through special data requests). Or, upon identification of problems and trends, pertinent material can be sent automatically to the appropriate organizations. Special data requests are a primary information transfer catalyst and are typically made with respect to a specific need. Overall, data and information can be retrieved to support topical areas such as reliability, maintainability, and safety analyses; maintenance planning; design improvement; manufacturing planning; and procurement decisions. Many potential applications exist.

A. Data Collection

Both product-specific experience and industry-wide generic data on similar parts should be collected. Experience data represents that obtained, for example, from developmental testing; manufacturing inspections, tests, and screens; field experience; and design analyses. Industry-wide generic data represents that obtained from existing data systems. Initially, a data system can be based on industry-wide data. However, when actual system and product experience data becomes available, it is incorporated into the data system as the primary resource.

The following are examples of hardware-oriented industry-wide data systems [3]. These data systems may provide a good source of initial material for input into the TMA data system, as well as support similar system and product analyses. Note that this type of information does not replace the need to acquire actual product or process experience data.

IEEE Reliability Data Manual (IEEE STD 500-1985)—A source of reliability data for electrical, electronic, and mechanical components. The data includes failure modes, failure rate ranges, and environmental factor information on generic components. Reliability data appears in the form of hourly and cyclic failure rates and failure mode information. Low, recommended, high, and maximum failure rates are given for each individual mode.

Government-Industry Data Exchange Program (GIDEP)—A cooperative program between government and industry for exchanging data to reduce the time and cost of researching relevant areas. The program provides a way to exchange certain types of technical data used in research, design development, prediction, and operation of systems and equipment used mainly in electronic or electromechanical application. Participants in GIDEP are provided with access to several major data interchanges: engineering data, metrology data, reliability-maintainability data, and failure experience data. The information is applicable for qualitative studies such as failure modes and effects analysis, decision tree analysis, and event tree analysis, as well as for quantitative studies such as reliability prediction, test interval calculation, or spare part studies.

Nonelectronic Parts Reliability Data Notebook (NPRD-3)—A publication providing test and reliability data primarily from military and space applications. The information is presented in four sections: (1) generic level failure rate data, (2) detailed part failure rate data, (3) part data from commercial applications, and (4) failure modes and mechanisms. The information is suitable for qualitative studies such as failure modes and effects analysis and decision tree analysis, as well as quantitative studies such as a reliability prediction of systems composed of nonelectronic parts.

MIL-HDBK-217 Reliability Prediction of Electronic Equipment—A compilation of data and failure rate models for nearly all electronic components, along with qualifying factors that design or reliability engineers can use to perform reliability analysis on electronic systems in specific applications. It provides a basis for performing, comparing, and evaluating reliability predictions of related or competing designs.

B. Data Processing

The objective of data processing is twofold. First, the system does all primary calculating, sorting, formatting, and organizing of the data and information to make it available for use as quickly and as accurately as possible. Second, the data banks (and files therein) are defined to aid potential users in readily gleaning pertinent and essential data.

This can be achieved through three fundamental data processing task elements: (1) preprocessing of data by participating organizations, (2) transfer of data to the centralized data system, and (3) specialized analysis of trends and significant problems highlighted by the data system.

It is common for user-descriptors to be defined with the input and output data. Examples include the following [4]:

1. Item Descriptors. A characterization of the components and products for which engineering data is being collected. Unique data elements are necessary for each product type to characterize it properly. For example, the data elements or parameters used to characterize and classify a crankshaft certainly will differ from those for an electronic microcircuit device. Item descriptors should be chosen for their known or postulated influence on reliability, quality, and/or safety.

2. Application Descriptors. A characterization of the possible reliability, quality, and safety influences relative to the manner in which the item being tracked is applied in the component and product. They help to classify the nature and level of application stresses that are present during product operation.

3. Operational Descriptors. A characterization of the environmental and operational stresses impacting a unit. Since individual product units are likely to be used in different geographic locations under differing load conditions, the unique operational conditions and environments to be experienced are recorded. For controlled tests, the specific test conditions are logged.

4. Operating Data. The actual operating record in appropriate units (hours, cycles, actuations) experienced by a unit during test or operation. Usually, the data is recorded over defined calendar periods or test duration. Where cyclic operation is involved, actual cycling rates, and/or on-and-off

periods are logged, since many units are susceptible to transient stresses, thermal expansion rate differences, nonuniform lubrication, and other detrimental influences that are not present during continuous operation.
5. Discrepancy and Failure Event Data. The discrepant and/or failure event. The event record should indicate the time of failure (and, if possible, the time of failure), how the event affected product function, a description of the event, and where possible, the failure mode and/or failure mechanisms. Also, the correction action taken to alleviate the discrepant event is important.

C. Data Reporting

There are three general ways in which data and information are retrieved from a data system: by written or telephone request, through periodic standard reports, and by direct access (via computer terminal). Also, a modem retrieval mechanism could be established at some point after the data system has matured.

Once the mechanism for data retrieval is in place, a broad range of analyses are possible using the basic data elements resident in the data system. Analyses could produce information for tracking reliability, safety, and quality status and progress for a specific product development/manufacturing program. Or, general engineering information could provided for application to future design and manufacturing improvement efforts.

Examples of output information include

- Identification of data recording periods and all discrepancies with reference to tests
- Failure rates (product, component, and part) and modes
- Reject rates by inspection and by screen tests
- Inspection, test, and screen effectiveness factors
- Mechanisms responsible for failure
- Recommended or accomplished corrective actions
- Failure analysis reports of hardware discrepancies, including accumulated operating hours to time of failure, failure modes, and cause and type of failure modes
- Cumulative points of failure events versus time
- General reliability analysis that correlates design analyses with test results and field experience

V. IMPLEMENTATION

The TMA data system will undoubtedly reside on a computer for ease of use and efficient data management. This provides an easy method for inputting and extracting data and information to assist in evaluating product performance throughout its life cycle. Such a system allows design, manufacturing, and industrial engineers; management; and other potential users to audit manufacturing system and product reliability, safety, and quality.

Extending a data system operation beyond a manual mode provides

1. Uniformity and standardization
2. Increased prediction/analysis speed
3. Reduced prediction/analysis costs
4. Automated compilation of output data
5. Compatibility with other corporate data systems

A computer-based, real-time system enables managers and engineers to develop an acute awareness of reliability, safety, and quality issues as they affect product and manufacturing improvement efforts. With a strong grasp of these issues, the ability to attain TMA is enhanced through optimization of management and engineering activities.

A well-structured TMA data system supported by failure analysis and corrective action also provides the means to maintain visibility over the effectiveness of the overall engineering and manufacturing programs. Product or system adjustments can then be made as necessary to minimize cost and maximize effectiveness.

As the data system evolves, other useful input data and information—aside from reliability, safety, and quality—will be identified. Similarly, more sophisticated statistical analysis methods can be incorporated, enabling a broader range of useful output intelligence to be produced in response to the changing needs resulting from technological advances and increased user awareness. In essence, the data system remains dynamic to accommodate advances in technology and processing capabilities.

VI. SUMMARY

This chapter addressed the establishment of a TMA-focused data system. The objective of such a data system is to provide a centralized depository for pertinent management, product, and manufacturing system data and information that are routinely generated. The availability of data and information reports allows for smarter management decisions based on specific status and trends and alerts concerning major problems. One of the biggest benefits comes from the ability to capture lessons learned for later reference.

The data system needs to be structured to easily capture key data and information for a vast array of topical areas including reliability, safety, and quality. Typically, two fundamental data-base categories are included in a comprehensive system: the technical information base (TIB) and the quantitative data base (QDB).

There are three key tasks in the operation of a data system: (1) data collection, (2) data processing, and (3) data reporting. These tasks enable the data and information to be recorded, transmitted, stored, and ultimately retrieved upon demand.

Packaged data systems are commercially available for all levels of computer technology. The data system is dynamic in its sophistication and contents. As the system matures, the scope of data captured should increase. Also, the technology used to support the system's operational requirements should continually keep pace with user demands for simplicity in data input and retrieval.

VII. QUESTIONS

1. Discuss the dangers of collecting all the data possible.
2. Discuss the need to maintain a centralized data system.
3. What type of program is necessary to support the data system?
4. Review Table 1. What additional data might a corporation's president require?
5. What are the two categories of data bases?
6. Identify and discuss various factors that would affect data system evolution.
7. What are the three major tasks involved in a data system operational cycle? What are the major milestones?
8. What are the purposes of generic data systems?
9. What are user-descriptors? How do they benefit data system users?
10. How does the data system interface with failure analysis and corrective action?

VIII. REFERENCES

1. Kelly, A.: *Maintenance Planning and Control.* Butterworths, London, 1987.
2. Cosenza, R. and Duane, D.: *Business Research for Decision Making.* PWS-Kent Publishing, Boston, 1988.
3. Brauer, D. and Bass, S.: *Reliability Analysis of Gas-Engine Heat Pump Systems: Methodology, Model, and Data System.* Gas Research Institute, Chicago, 1986.
4. Cleveland, E., and Duphily, R.: *Power Plant Data Systems.* Electronic Power Research Institute, Palo Alto, California, 1978.

9

Data Recording and Feedback

In support of TMA it is necessary to collect pertinent data and information and then use it to improve corporate management, engineering, and the overall manufacturing system. This requires having a disciplined, dedicated data recording and feedback program in place. The technology to perform the data collection must be included in such a program.

This chapter consists of two sections. The first addresses corporate data communications, including the common types of technology used to collect data. The second describes the mechanics of a failure recording, analysis, and corrective action system to support timely and meaningful data recording and feedback.

I. INTRODUCTION

The key to manufacturing control, as well as corporate control, is the ability to know what is happening. This is why a TMA-focused data system is established, as discussed in Chapter 8. To collect data that provides knowledge about the overall corporation requires a data recording and feedback system (DRFS). Ideally, this is a real-time activity that facilitates immediate and knowledgeable management, engineering, and operator responses to immediate situations.

A real-time scenario eliminates the poor practice of addressing problems a week or two after they occur. It does little good to have a DRFS in place if it is difficult to collect, digest, interpret, and respond to the manufacturing and product information that is generated.

The need for a DRFS is inevitable. The vast amounts of information generated by a corporation every day provide valuable insight and guidance for attaining TMA. The key is to maintain an efficient corporate data base; that is, to collect only information that is useful. Data may be useful in either the short term or the long term.

It must be recognized, however, that it is impossible to collect all the information that is generated. Nor should we try. We must remain focused on data-base efficiency. This does not imply passing over information to limit data-base size. It is always easier to reduce the amount of data collected and maintained than it is to add collection and maintenance requirements later.

A DRFS typically has one of two main orientations. In the first, information (for example, a document) is not replaced, but only managed. This is analogous to a card catalog system. Important information is collected and maintained, but not every possible piece of information goes into the system. In the second orientation, the system contains as much information as possible. For this alternative to work, there must be a comprehensive management practice, because the information and the DRFS itself become highly complex and difficult to use.

Regardless of which orientation the system uses, it is necessary to efficiently record the data and ultimately return it to the potential user. The best way to do this is to implement a real-time, automated data collection system. Many commercially available systems of this type are now commonly used. Such systems are referred to as *factory information systems.*

Factory information systems allow data to be collected and processed in real time, which provides great benefits in the manufacturing environment. In addition, applications of real-time DRFS are usually extended to corporate management, product development, and product deployment, as well as other activities.

A key benefit derived from factory information systems is increased management awareness, engineering efficiency, and operator productivity. This is achieved through a reduction in strain and frustration.

Chapter 8 addressed the fundamentals of data system planning. This chapter focuses on effectively capturing data and using it to continually improve the manufacturing system and its products and general corporate health. An overview is provided for several data recording media. Also, special attention is given to the need for a failure recording, analysis, and corrective action (FRACA) process as a means for feeding product information back into the continuous manufacturing system improvement initiative.

II. CORPORATE DATA COMMUNICATIONS

Perhaps the most important aspect of establishing a DRFS is to identify the type of data that is worth collecting and reviewing. Information is generated in four primary areas: (1) product, (2) manufacturing system, (3) corporate business structure, and (4) product operational environment.

As stated earlier, the potential amounts of acquired information are enormous. It is probably fair to say that data collection is one of the most time-consuming functions within a corporation. A critical element of company success, however, is the ability to acquire accurate and timely data and to ensure that it is fed back to all concerned business areas as soon as possible to enhance decision making.

Historically, DRFSs have one thing in common: paper. For the most part, information is recorded and fed back on paper. Get rid of all the reams of paper, and a major bottleneck in using manufacturing and product information is eliminated.

The next logical step is to embrace the concept of the paperless corporate operating environment. The technology exists to move in this direction, and the money saved in paper and handling costs alone would justify the hardware costs for the change.

This technology is commonly referred to as electronic shop documentation (ESD). Basically, computers are used to replace paper documents with electronic ones. A typical ESD system consists of a network of computers, operators's terminals, and software. Generally, any kind of paper document is adaptable to the electronic format.

Corporate success is enhanced by a data communications system that simultaneously handles product design, manufacturing, testing, and business area information. With fast, accurate communication among management, engineering, and the shop floor, the ability to attain and maintain TMA is increased. For example, flexible manufacturing cells can be reconfigured quickly to manufacture an alternative product without losing TMA. Good communication makes this continuity possible, since TMA lessons learned are recognized and are available as a resource.

Data communications can be based on one of three types of DRFSs: (1) manual, (2) semiautomated, and (3) real-time, automated. In the manual system, all information is manually recorded and stored on paper. This type of system is inefficient because mounds of paper rapidly develop. Such a system is inherently cumbersome and expensive.

The semiautomated system is similar to the manual system except that it takes advantage of computer technology. The computer generates paper documents and can maintain a data system. Computational errors are reduced, and the speed and diversity of the reporting mechanism is improved. However, this type of system is not real-time oriented.

The third type of DRFS is real-time, automated. This system permits information to be collected, verified, and processed as it is generated. The result is direct two-way communications between the computer and the user(s) of the information. A real-time, fully automated DRFS consists of engineering and data-base management software and a hardware network. The system integrates nicely into the corporate routine and facilitates simultaneous engineering.

As discussed in Chapter 3, simultaneous engineering is a very cost-effective approach to developing a product. Essentially, it works by getting contributions from applicable corporate business areas early in the development cycle. This results in a shorter overall product introduction time and a more robust product design.

With this in mind, data bases must be flexible enough to accommodate all possible data inputs and requests. The data base must be designed to accomodate the real world, not just some abstract data. This includes providing means for moving and manipulating data. Additionally, a data base must be accessible to all potential users through an on-line network capability. Poor accessibility does nobody any good and only forces redundant work. If the same information reports are generated for several departments, the benefits of simultaneous engineering are lost.

Automating the DRFS requires the use of information management programs, search programs, and/or form generation programs. Advances in state-of-the-art workstation technology increase the potential uses of DRFSs. Workstations create forms, call up and run subroutines or outside programs, access other systems, and search and find data based on ambiguous instructions.

A. Types of Databases

An important element of an automated DRFS is its supporting data-base structure for storing information. Numerous data-base management systems are commercially available that provide a way to develop custom data-bases quickly. This allows the development of data-bases for managing large amounts of distinct information. The key to a successful data-base is its ability to provide the right information for each specific request.

Two data-base structure approaches are commonly ussed. The first approach is the *relational data base*. Relational data bases treat data as an enormous table in which each record is represented as a row and the attributes as columns. This provides two major advantages: (1) it is easy to ask for data sorts, and (2) sorts can be made to address several distinct data files. This provides great flexibility in defining data sorts that can identify all the information available.

The other common approach is the *object-oriented data base.* In this type, each record is treated as a separate object with a collection of attributes. These objects are automatically incorporated into a larger data assembly.

B. Automatic Identification Technology

Recording information is an integral element of DRFS. Doing this efficiently is desirable. A principal way to achieve this is by using some form of automatic identification (AID) technology [1].

AID has a broad application horizon. It serves as a supporting technology to computer-integrated manufacturing (CIM) (discussed in Chapter 5). It also is a valuable tool in controlling inventory processes, keeping track of personnel information, and ensuring corporate security. Benefits include lower operating costs through higher productivity, better inventroy cotnrol, and improved customer service. This technology ensures accurate and efficient collection of data in response to corporate needs.

A corporation's use of AID technology enables it to monitor itself in real time, and more closely. For example, the effectiveness of material flow control can be greatly enhanced which fulfills a baseline requirement of CIM systems.

Tracking ability is an inherent feature of AID technology. Everything can be tracked and monitored, from work in process to finished production units. In some cases, people are tracked for human resource purposes. AID technology is applicable to a number of diverse environments within the corporation.

The fact that AID generally can be applied wherever automatic recording of information is appropriate means greater productivity in performing the information recording activity. AID has eliminated the need for a manual recording system. By eliminating the human element, information accuracy increases, since fewer errors will occur during recording.

Many specific technology types comprise the AID family. Each has the goal of rapid and accurate data collection and recording, but differs in its approach to capturing and processing data. Bar coding is the most popular form of AID in terms of size and growth rate [2]. However, other technologies exist such as optical character recognition, voice entry, vision, magnetic stripe, and radio frequency. Each provides advantages, depending on the specific application.

1. Bar Coding

Bar coding refers to the use of a symbol that corresponds to combinations of digits, letters, or punctuation marks. It consists of a pattern or narrow and wide bars and spaces to represent information. Each pattern represents a different alphanumeric character. The technology integrates four independent processes: (1) printing, (2) reading, and (3) transforming the code to a time phase (which is then decoded), and (4) a coding scheme. There are many commercial systems available; however, each is typically unique, making it difficult to compare system specifications.

The most common bar code systems use scanners that read bar code information by shinning a light at the symbol and monitoring the amount of light reflected back. The light source is an incandescent light bulb or a light-emitting diode (infrared or laser). The reflected patterns of light are captured by a photodetector and ultimately converted into digital signals and transferred to the data system.

A second type of scanner uses a solid-state light-sensitive element, or charge-coupled device. With this type, the reflected image of the bar code information passes through an ordinary camera lens and is then projected onto the charge-coupled device.

Bar codes provide a way to record data rapidly. With a single scanner pass, data is read and input to the data base at a speed many times faster than that possible by manual methods. Bar codes are also convenient. They can be read at a distance, they provide unique sensing capabilities for timing purposes, and they are very flexible and long lasting in monitoring manufacturing systems.

Bar codes reduce the potential for data recording errors because they are reliable and user-friendly. They work well with shop personnel who have a low skill level or understanding of the corporation's manufacturing system.

2. Optical Character Recognition

Optical character recognition (OCR) is similar to bar code technology in that it uses a scanning technique to record data. OCR reads alphanumeric characters, not coded information. The two types of OCR systems used are document carriers and page readers. The latter reads a full-size page, while the first reads only a few lines of information at a time. The use of this technology is reduced in manufacturing systems where rapid data recording and data accuracy are required. OCR use is expected to increase with increased use of vision systems.

3. Voice Entry

Voice-entry technology is based on pattern recognition, as is bar coding. Instead of images, however, words of a preprogrammed vocabulary are recognized. The user speaks into a microphone and the phrase is recognized by the machine. The information is then transformed into electrical signals that are subsequently sent to the data system.

Depending on the sophistication of the system, the recorded information is interpreted and related information or instructions are fed back to the user. A key part of the system enables users to teach the system to recognize their voices in the environment in which they will be working. This allows background noise to be essentially washed out. The advantage of this technology is that the user's hands and eyes are free to perform a manufacturing activity.

4. Radio Frequency

Radio frequency (RF) technology is typically used in dirty and harsh manufacturing environments. This includes layout situations where physical obstructions or manufacturing system characteristics deter the use of other data recording media. RF frequently replaces bar codes or magnetic stripes where they are rendered useless due to reading difficulties.

The technology uses bidirectional radio signals to transmit data between an RF reader and a transponder located on, in, or near the object to be detected. The transponder operates on a specified frequency and transmits its message when it is addressed by a reader.

Radio frequency is commonly used in automation and material handling applications where there is no line of sight between scanner and identification tag or where read/write capability is required. A transponder on the object being tracked sends a unique signature or data stream when addressed by the transmitter/reader. The antenna picks up the signal, and the reader decodes and validates the signal for transmission to the computer containing the data base. This technology also enables other AID technologies to transmit information to the data system via a remote radio link.

5. Vision

Machine vision is very complex and consists of a television camera, a monitor, a keyboard, and a processor. A camera picks up the image and sends it to the processor as an analog signal. The processor converts the signal to a digital matrix and compares the picture with information stored in its memory. Upon a match, the image is recognized and an output is generated. This may consist of a series of identification numbers, a numeric representation of the image, a list of flaws, or a go/no-go command.

In general, any legible (to a person) characters provided on a label can be read and interpreted by a vision system. Characters or objects can be recognized at all angles, including upside down, and different lenses provide changes in the observable depth of field. The speeds at which characters and objects are recognized varies. Generally, character recognition ranges from thirty to sixty items per second. The speed of recognizing objects changes based on the number of features to be recognized or inspected, as well as the environmental conditions.

6. Magnetic Stripe

Magnetic stripe technology involves encoding information onto a special material. A decoder reads the magnetic stripe and passes the information on to a computer. The advantage of this technology is that large amounts of data can be stored easily on a single stripe. Therefore, it is widely used for personnel identification badges and bank credit cards.

The amount of information that can be on magnetic stripes is significantly greater than that possible in bar codes. Magnetic stripes also hold up to wear and tear. Even if a magnetic stripe is mutilated or subjected to heat, dirt, grease, and other factory conditions, it generally still can be read with a high degree of reliability.

C. Programmable Logic Controller

The previous subsection described several technologies that are used essentially to capture and transfer data a step or two away from a process. They do not directly influence any ongoing processes by making system operational adjustments. To do this, the programmable logic controller (PLC) is used.

PLCs have been around for a long time and are, in recent years, growing rapidly in their potential for monitoring and controlling manufacturing systems (and processes supported thereby) in real time. They are well known for their efficient and flexible industrial control capabilities; the current trend is for them to be exploited increasingly for their data gathering ability.

State-of-the-art PLCs are built with advanced instruction sets, memory, data processing, and communication capabilities. Consequently, PLCs are used to perform data acquisition and analysis in addition to other traditional control duties.

PLC technology consists of microprocessor-based central processing unit, user memory, inputs and outputs, and a power supply. PLCs control machines and processes by accepting data from field input devices, using the data to solve user-written control programs, and creating outputs. The nature of the PLC's memory allows quick changes to be made to the control program.

The maturity of networking and communication capabilities is allowing PLCs to interface with host computers, terminals, displays, other PLCs, and other devices such as expert systems [3]. This gives operators, as well as the DRFS, immediate access to vital process information needed for analysis and decision making.

There are two primary reasons to use PLCs for data acquisition. First, PLCs are widely used in an automation role. The expansion of the PLC's role to include data acquisition has reduced the need to purchase additional or dedicated data acquisition devices. Second, PLCs have direct access to process and systems information.

In a data acquisition role, PLCs can monitor, collect, and store manufacturing data. This improves the overall productivity and efficiency of computer-controlled manufacturing systems. In addition, material waste is reduced and product quality improves.

An advantage of the PLC is its ability to monitor a broad range of manufacturing variables and immediately manipulate the data for analysis. Examples include calculating material misfeed rates, the number of "good" versus "bad" parts over time, and system running efficiency. This information can be displayed in real time for immediate interpretation by an operator.

An additional benefit of the PLC mathematical capabilities is that there is no need for shop personnel to worry about statistics and calculations. Because the PLC collects vast amounts of data by itself, it is a significant contributor to increased efficiency and productivity.

III. FAILURE RECORDING, ANALYSIS, AND CORRECTIVE ACTION

If the DRFS does nothing else, it must provide a mechanism for failure recording, analysis, and corrective action (FRACA) [4]. A formal FRACA effort allows the early elimination of failure causes and forces the reliability of manufacturing systems and products to grow. A natural extension of this increased reliability is the attainment of high levels of operational availability.

Although FRACA is frequently emphasized as an activity employed for hardware items, it need not be limited to that. Nonhardware items such as late shipments, faulty invoices, and so on, are also serious problems or failures Keep in mind that wherever a word referring to hardware appears, it can most likely be exchanged for any product going out from a staff function.

To get the full benefit of FRACA, it must be implemented early during the design/development phase of a manufacturing system or product. This allows failure causes to be effectively identified at the point where it is most cost-effective to implement corrective action. As designs, documentation, and hardware mature, corrective action is still identified, but its implementation becomes more difficult and costly. This underlines the importance of having a closed-loop FRACA activity in place at the beginning of a project.

The term *closed loop* refers to an activity that begins with the source of the reported problem and ends with reporting the results of the failure analysis and corrective action back to that originating source. To make this closed loop work in a timely manner, FRACA is integrated into the existing TMA data system and DRFS.

The overall effectiveness of a FRACA process depends on the accuracy and thoroughness of the input data; that is, reports documenting failure or malfunctions and failure analysis. The essential data inputs are made by a failure reporting activity. The scope of failure data includes all information pertinent to the failure, which facilitates determining the failure cause.

A prerequisite of a healthy FRACA process is the use of a standardized failure report. The report itself must be designed to simplify the documentation of failure data and enhance the effectiveness of the FRACA process. This means having a form that is easy to fill out and captures only the essential data.

The FRACA report merely initiates the closed-loop process. As shown in Figure 1, the report follows a rigorous path. This includes performing a correlation/trend analysis in the data system, performing a detailed future analysis, determining corrective action, receiving concurrence from the failure review board, and finally, implementing the corrective action.

Keep in mind that FRACA is only one piece of the whole DRFS, but a very important piece. Also, the data system itself must be designed to support the DRFS. As discussed in Chapter 8, the data system needs to be robust in its content, as well as functionally able to perform various statistical, correlation, and trend analyses.

A key objective of the FRACA process is to routinely identify failure trends and correlations as they become evident. If the data system indicates the appearance of a trend, a failure analysis is performed. This serves as a means for determining the failure cause and the appropriate corrective action needed to eliminate, or minimize, its recurrence.

Failure analysis may be as simple as technical dialogue between design and reliability engineers to identify the failure cause(s). In some cases, however, formal, detailed laboratory failure analysis may be required to reveal the failure mechanisms and provide a strategy for deriving an effective corrective action. Note that corrective action generally focuses on design changes, process changes, or maintenance changes. These options exist for either manufacturing system or product problems.

The FRACA effort needs to be flexible enough to accommodate all failure occurrences during a project, including those that occur during actual operation. Along with this flexibility, a key FRACA output is a summary report that delineates information about failures and evidence of trends, as well as the extent of contemplated corrective action and its estimated impact (cost and effectiveness).

A. Key Elements of the FRACA System

As stated earlier, FRACA is a closed-loop, real-time activity for identifying failure/problem areas. This naturally leads to the initiation of aggressive follow-up activities to implement and verify corrective action. Effectively correcting observed problems requires rigorous and persistent data tracking and coordination of all the necessary actions performed by different organizations and disciplines.

To have a cohesive effect, a FRACA plan is developed. The plan identifies all the provisions needed to ensure that proper data analysis is performed and

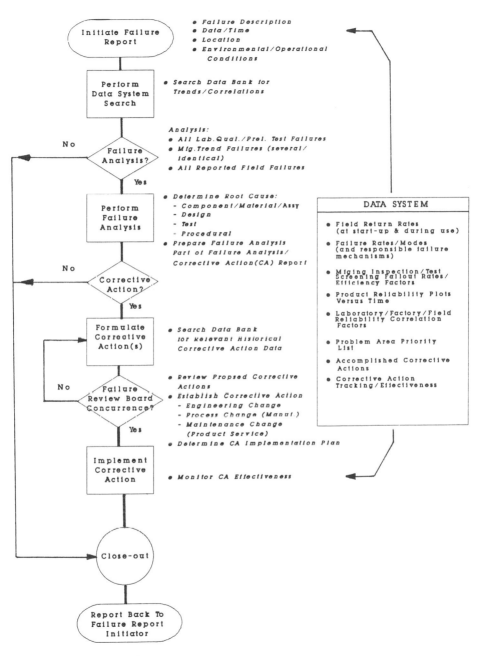

Figure 1 FRACA Interaction with the Data System

that effective corrective actions are taken on a timely basis. Additionally, the various levels of assembly are identified, definitions for failure cause categories are provided, logistic support requirements are identified, and the FRACA data items required are identified.

B. Responsibility

In most cases, the reliability engineering department (or group performing this function) is responsible for instituting and managing the FRACA activity. It establishes policy, provides direction, and monitors the status of a failure analysis investigation. Specific responsibilities include

1. Assigning identification numbers to reports received, completing the reports, and determining the need for failure analysis and correction action
2. Performing trend/correlation searches in the data system
3. Conducting failure analyses and corrective action investigations
4. Monitoring the effectiveness of the corrective actions implemented
5. Informing the appropriate corporate departments, as well as the report originator, of problems closed out, including the results of the investigation and the action taken
6. Maintaining a FRACA experience base as an integral part of the overall data system

C. Activity Architecture

Figure 2 depicts the FRACA activity as a closed-loop operation. This configuration consists of several fundamental steps that must exist for the activity to be complete and effective.

The closed-loop activity in its leanest form consists of the following steps:

1. Observe and document the problem.
2. Verify that the problem actually occurred.
3. Isolate the problem to the lowest level possible.
4. Implement interim corrective action.
5a. Determine the problem cause (failure analysis).
5b. Search the data system for trends/correlations.
6. Determine and approve the necessary corrective action encompassing design change, process change, or procedure change.
7. Implement the corrective action and monitor its effectiveness.
8. Close out the problem report.
9. Feed back the results of investigation to the originator of the problem report.

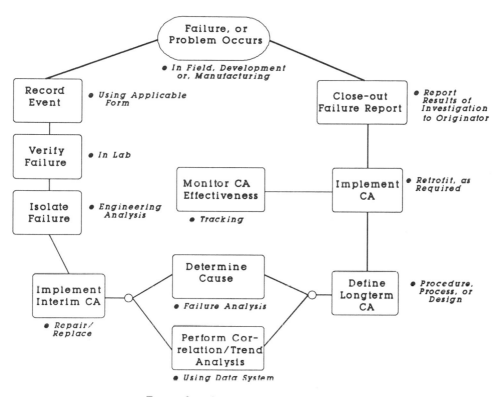

Figure 2   Closed-Loop FRACA Activity

D. Reporting

Obviously, to make this activity work, it is imperative that failure reporting and resulting corrective actions be documented effectively. FRACA forms must be designed to meet the needs of the individual manufacturing system or product, as well as departmental responsibilities, requirement, and constraints.

Key information to capture includes part identification data, conditions under which the problem occurred, operating parameters indicating degradation, replacement part(s), repair times, references to applicable plans and procedures, and complete details leading up to or surrounding the incident. Failure analysis information includes a complete description of the failure parts, including manufacturer and lot production code. Once the failure analysis is complete and recorded, the decision is made whether or not to proceed with corrective action. Finally, the problem cause is described in detail and recommendations are made for corrective action.

IV. SUMMARY

This chapter addressed why establishing and maintaining a corporate-wide data recording and feedback system (DRFS) is desirable. A DRFS enables all persons involved in the manufacturing effort to feed the TMA data system with timely and accurate data. The ideal situation is to have a common data base, a resource shared in real time, on-line and updated immediately when new information is generated.

Numerous automated technologies are available to collect data in real time. These are referred to as automated identification (AID) technologies and include bar code, optical character recognition, voice entry, radio frequency, vision, and magnetic stripe. Also, programmable logic controllers are widely used and are very capable data collectors.

An essential part of the DRFS is a working closed-loop failure recording, analysis, and corrective action (FRACA) activity. This activity ensures that failures are accurately reported and thoroughly analyzed and that effective corrective actions are taken on a timely basis to eliminate, or minimize, recurrence of the failures. The FRACA activity is most cost-effective when it is open to all failure and problem data regarding both hardware and non-hardware output products and systems.

V. QUESTIONS

1. What is meant by real-time data collection?
2. Discuss the various types of data that different departments might require in real time.
3. What are the major benefits of moving to electronic shop documentation?
4. Discuss the disadvantages of a manual DRFS.
5. What are the two types of data bases? Determine what type of data base you access.
6. What are the major technologies used for automated identification? Identify applications of each in the manufacturing system.
7. Discuss the relationship between FRACA and the TMA data system.
8. What is a closed-loop FRACA process?
9. In the FRACA process, what is the difference between short-term and long-term corrective action?
10. What type of information is necessary to close out a failure report?

VI. REFERENCES

1. *Barcode Basics*. Allen-Bradley Company, 1989.
2. Automated Data Collection. *Production*, April, 1989.
3. Hayes-Roth, F. (editor): *Building Expert System*. Addison-Wesley Publishing Company, London, 1983.
4. *RADC Reliability Engineer's Toolkit: An Application Oriented Guide for the Practicing Reliability Engineer*. Rome Air Development Center, Rome, New York, 1988.

Index